Preface

I wrote this book to first show how a learning disability, such as dyslexia, can effect creative thinking. Since Einstein displayed a learning disability in his early years, many modern educators believe he had dyslexia.

Secondly, I wrote this book to show how I, as a diagnosed dyslexic, have similar thought processes as Einstein, even if it is at a different level.

Thirdly, I try to show how I use modern theories (i.e. quantum theory and M theory) to explain current mysteries of theoretical physics.

I do not or cannot use higher math for these findings. I simply rely on "end results" of theoretical physics for my base. Einstein said he was a poor mathematician, but compared to him I fall way short!

The conclusions that I draw from current theories leads me to postulate and theorize about: dark matter, dark energy, super-symmetry, the eleven dimensions of space-time, the unification of forces, the spin of particles, electron orbital's, eternity, the theory of everything (TOE, as the "Grand Original Design" {G.O.D.}), beginning and end of the universe, plus I introduce a new particle, --- The Space-Time-Energy-Particle (STEP).

I postulate that quantum theory (and as Einstein believed) is incomplete and "what can happen-will happen" is only true if STEPs entangled with quantum theory allow these happenings to take place.

I also formulate a Holy Grail Equation for All of Physics, which I term "Stauffer's Holy Grail Equation" to show ownership and diversity from other attempts for a one inch equation of the entire Observable Universe (OU).

This is THE NEW REALITY.

Dedication

This writing is dedicated to my niece and nephew Debra Pleace and William Laurence Stauffer III who both left us way too early in their life!

Their voluntary departure by way of self-inflected means with a shot gun and a river bridge is so sad.

Life is precious and as great as Eternity is, a Soul gets nourishment in our universe and should be cherished to the n'th degree!

Also to Coach Foster at Harmony High School and to all his Special needs Buddies he fosters with pure dedication and sincerity.

Below are some of my new and established definitions that relate to the paradigm shift for a" NEW REALITY".

Time, Space, Spacetime.......January 30, 3:41 AM, 2016 ADAM=at morning AD=after death

Time: denotes the temporal tesseract exchange of two Planck size dark matter cubic manifolds (with one being a Calabi Yau on a three sphere inscribed within the other) at every Planck tic (10^{-43} seconds) of time.

Space: denotes the spatial bounties of the two temporal manifolds exchanging position by tesseract rotation.

Spacetime: denotes the four dimensional harmonic relationship of space and time with six spatiotemporal additional dimensions depicted as worm holes for concentrated energy (normal matter) to move through. Spacetime is independent of quantum string matter since it is only the background in which matter travels through as traveling waves. The eleventh dimension is the onion skin between the manifolds that house the 2-spin O-ring graviton. This allows the observable universe (OU) to be smooth (low energy) because the graviton is only felt in the OU when it is at the equator of the three sphere. This allows quantum theory to mesh with the general theory of relativity.

Dark matter: denotes concentrated dark energy.

Dark energy: denotes the dilutions of dark matter such that it cannot be detected as additional matter in the observable universe (OU), much like when a photon doze not have enough energy to extract an electron from its orbit.

Space-time-energy-particles (STEP): denote the quanta nature of the OU in which there are approximately ~$10^{183+/-3}$. Each particle is a Planck cubic in size and contains the eleven dimensions. All are syhronized with each other *which accounts for entanglement of string matter through time channels void of space passage.*

The Holy Grail Equation of Physics: $Ef = (Dm)(\Lambda a)$, where Ef = eternal force (also called the super force), Dm = dark matter, Λ_a = acceleration of the cosmological constant. All mathematical and Physics expressions must fit into this relationship of terms for this equation to explain the theory of everything, ToE.

The British Dictionary defines Spatial as:

1. Of or relating to space.
2. Existing or happening to space.

Temporal is defined as:

1. Of, relating to, or limited by time: a temporal dimension; temporal with spatial boundaries.
2. Lasting only for a time: not eternal; our temporal existence.

Spatiotemporal and Spatiotemporal patterns:

1. Belonging to space and time or to Space-time.
2. Complexity of spatiotemporal patterns can only be recognized over time. Any kind of traveling wave is a good example of a spatiotemporal pattern.

The Grand Original Design Tree for the ToE

Grand Original Design

| ToE

Holy Grail Equation of Physics

|

Complete theory of quantum mechanics

|

Space Time Energy Particles (STEPs)

| |

Standard model of Cosmology   Standard model of Particles

| |

General Relativity   Incomplete quantum theory

| | | |

Special Relativity   Classic Newton   Wave equation   Quanta

Eleven Dimensions of Space-Time

(3-spatial / 1-temporal and 7 spatiotemporal wave patterns)

Albert Einstein, along with Hermann Minkowski, taught that four dimensions of volume are needed to explain special and general relativity along with Space-time. Others felt that up to six additional dimensions are needed to define Space-time dimensions. It was not until Dr. Edward Witten improved the string theory by adding an eleventh dimension (the M theory or superstring theory) that physicists saw a real workable framework to develop a theory of everything (ToE).The non-string physicists also needed eleven dimensions to make their Loop Quantum Gravity (LQG) theory fit into their equations. In order for the M superstring theory to be compatible with the LQG theory there must exist a smooth independent space-time background for gravity to function without compromise. Six of the string theory and M theory dimensions are often described as a six dimensional Calabi Yau Manifold. The eleventh dimension of the M theory developed by Dr. Witten relates to gravity and gravity is the one force that impedes the unification of the two theories by way of low energy requirements. When the Observable Universe (OU) is considered a collection of Planck size sphere-like hidden manifolds, a pattern is created that incorporated the eleven dimensions at the Planck scale. The array of possible models (10^{500}) of three spatial dimensions (height, depth and length) plus the temporal dimension of time and the remaining dimensions (5-11) are referred to as the string theory landscape. This landscape is considered rough as opposed to smooth, which presents a problem when trying to combine the two theories! Recently Dr. Kenneth R. Stauffer has proposed a theory that incorporated temporal, spatial and spatiotemporal pattern-parameters to all of the eleven dimensions to solve this problem. But first sound definitions of spatial, temporal, spatiotemporal patterns, Space-time and time need to be addressed by definition.

The British Dictionary defines Spatial as:

3. Of or relating to space.
4. Existing or happening to space.

Temporal is defined as:

3. Of relating to, or limited by time: a temporal dimension; temporal with spatial boundaries.
4. Lasting only for a time: not eternal; our temporal existence.

Spatiotemporal and Spatiotemporal patterns;

3. Belonging to space and time or to Space-time.
4. Complexity of spatiotemporal patterns can only be recognized over time. Any kind of traveling wave is a good example of a spatiotemporal pattern

Space-time

- "Also called Space-time continuum. The four-dimensional continuum, having three spatial coordinates and one temporal coordinate, in which all physical quantities maybe located.
- The physical reality that exists within this four-dimensional continuum.
- Of, relating to, or noting a system with three spatial coordinates and one temporal coordinate."

Time

"A temporal dimension is a dimension of time. Time is often referred to as the "fourth dimension" for this reason, but that is not to imply that it is a spatial dimension. A temporal dimension is one way to measure physical change. It is perceived differently from the three spatial dimensions in that there is only one of it, and that we cannot move freely in time but subjectively move in one direction.

The equations used in physics to model relativity do not treat time in the same way that humans commonly perceive it.

The equations of classical mechanical are symmetric with respect to time, and equations of quantum mechanics are typically symmetric if both time and other quantities (such as change and parity) are reversed. In these models, the perception of time flowing in one direction is an artifact of the laws of thermodynamics (we perceive time as flowing in the direction of increasing entropy)"

Stauffer developed the 11 spatial/temporal / spatiotemporal dimensions in the following manor:

- Planck time flows at the speed of a Planck's tic and at a spatial distance of a Planck's length.
- Time flows in the direction of increased entropy.
- Space is a two dimensional spatial p-brane (1st and 2nd dimensions) and at the Planck scale it is one Planck's length squared.[22]
- Time is not spatial. It does not happen in space. It is a temporal dimension with spatial space boundaries only when an elementary particle passes through one space d-brane to another dose time record. When a string elementary particle is in the temporal region it can travel super-luminously to a spatial p-brain and will be recorded as one Planck of time. This is how "spooky action at a distance," extra sensory perception (ESP) and prayer can be accounted for!
- Space-time continuum. The eleven-dimensional continuum, having three spatial coordinates, one temporal coordinate and seven spatiotemporal coordinates, has physical connotations. Others referring to the M theory consider the eleven dimensions to be 10 spatial and 1 temporal.
- Space-time is the physical reality that exists within this eleven-dimensional continuum.
- Space-Time-Energy-Particle (STEP) relates to, or noting a system with three spatial coordinates, one temporal coordinate and seven spatiotemporal coordinates at the Planck scale level.
- The speed of light (Planck spatial length/ Planck temporal tic of time) are the 3rd and 4th dimensions.

- A STEP is Dark matter (defined as concentrated Dark energy) acting against its self as Zero Point energy in the confined volume of a space-time-energy-particle (STEP).
- The STEP has two alternating forces that are described as a Planck cube exchanging location with a Planck cube on a 3-sphere. The 3-sphere is considered a Calabi yau tesseract that has six spatiotemporal worm holes for elementary string particles to navigate through, such as;
  - An Axillary wormhole that is "0" radius units from the axis of the 3-sphere. (5th dimension) Note; it is the axis.
  - Two cork screw wormholes that have a diameter that is equal to "1/2"of the radius of the axis of the 3-sphere. (6th and 7th dimensions)
  - Two cork screw wormholes with diameters equal to the 1 radius from the 3-sphere axis. (8th and 9th dimensions).
  - A ring worm that has a diameter that is 3/2 radius of the 3-sphere axis (10th dimension)
  - The interior volume that the ½, 1, 3/2 and 2 radiuses is where interactions take place for all elementary string particles.
- The 11th dimension is also a spatiotemporal onion skin region separating the two alternating Calabi Yau tesseracts. The onion skin has a diameter of a 2 radius and is where the graviton operates in this region. The onion skin separates the two tesseracts. When it (the graviton) is fully stretched at the equator of the tesseract Calabi yau sphere, its maximum strength is felt in the (OU). This is why gravity seems weaker than other forces of the OU, since when the graviton is not fully extended it is not in our universe.
- Space-time is considered smooth due to its non-grainy geometry (a cube alternating with a cube on a 3-sphere with tesseract rotation). The synchronized nature of the STEPs and their wormhole passageways through the tesseract-Calabi yau 3-sphere, where elementary string particles move as traveling waves, have a non-grainy geometry. The graviton also operates in a smooth region of space-time (between the two cubes depicted as an onion skin).

The 5th through 11th dimensions are considered to be spatiotemporal as opposed to spatial or temporal for the following reasons:

- Their dimensions have length (spatial) associated with them, such as the distance of worm holes from the axis of the tesseract Calabi yau on a 3-sphere or the distance of worm holes from one p-brane to another.
- Their dimensions also have time (temporal) associated with them such as the Planck tic of time performed during the exchange of position every 10^{-43} seconds. The super luminal speed of the time corridor is another example.

If Stauffer's postulates bear out, then Space-time, as we know it will have added new insights with a complete working model of the

ToE, such as:

- An elementary string particle can move superluminal through time but only travel at the speed of light or slower when traveling through space-time.
- Time has only one direction with respect to space. It allows several pathways for specific elementary string particles, antiparticles, and sparticles to navigate.
- Time does not happen in Space, it happens with Space.
- Time is the Zero Point Dark energy and/or the Zero Pont Dark matter movement confined in a STEP where a Planck's tesseract cube is exchanging places with a Planck's tesseract Calabi yau cube on a three sphere every Planck's tic of time. (10^{-44} seconds).
- The definition of Space-time should be broadened by :

  1. Incorporating a total of seven spatiotemporal dependent dimensions where dark matter and/or dark energy are trading places with in a STEP, which allows elementary particles to travel in there accompanied wormholes, onion skin or ringworm passage ways.
  2. Space-time is the physical reality that exists within the eleven-dimensional continuum, of, relating to, or noting a system with three spatial coordinates, one temporal coordinate and seven spatiotemporal dimensional continuum coordinates.

3. The eleventh dimension (an onion skin around the tesseract/Calabi yau on a three sphere) houses the graviton 0-ring

By virtue of having a 3-sphere geometry alternating (by tesseractic exchange) with a cubic tesseract, smooth Space-time is obtained. This geometry results in a Loop Quantum Gravity (LQG) independent background. Background-independency allows general relativity to achieve a low-energy approximation.This allows Loop Quantum Gravity to achieve the goal of independence by operating in a smooth fabric of space-time in the general relativity frame work. Also, the smooth nature of the onion skin is critical for a low energy requirement. The synchronization between STEPs of space-time yield a background-independent way of looking at gravity in the micro and macro world. This allows a complete theory of everything (ToE) where the M theory and Loop Quantum Gravity are mutually compatible.

Acknowledgments

My love and thanks go out to Sharon Louise Mac Hose Ebling for her unwavering support and editing of this book. Also author Ric Bauer, Ryan Ebling and Lydia Martinez were instrumental in leading me to the right people to advance my work.

I also want to especially thank AnnMarie Ebling for late editing of my book. She usually gets disgruntled when she finds mistakes in any of the many books she reads, but this time she got to correct what she saw was wrong.

I would like to acknowledge Helen Cousins who did a final scan of my manuscripts for content and miscues.

Lastly, by the Grace of God, I acknowledge the Divine Intervention that occurred during the wee hours of the night when my creative thoughts came to fruition.

Nebula/IC434

About the image on the front cover as a quote from NOAO/AURA/NSF

"This exceptional image of the Horsehead nebula was taken at the National Science Foundation's 0.9-meter telescope on Kitt Peak with the NOAO Mosaic CCD camera. Located in the constellation of Orion, the Hunter, the Horsehead is part of a dense cloud of gas in front of an active star-forming nebula known as IC434. The nebulosity of the Horsehead is believed to be excited by the bright star Sigma Orion, which is located above the top of the image. Just off the left side of the image is the bright star Zeta Orionis, which is the easternmost of the three stars that form Orion's belt. Zeta Orionis is a foreground star, and is not related to the nebula. The streaks in the nebulosity that extend above the Horsehead are likely due to magnetic fields within the nebula. Close study reveals that many more stars are visible in the top half of the image. Stars in the lower half of the image are obscured by a dark cloud of hydrogen gas. The edge of this large cloud is the horizontal strip of glowing gas that bisects the image. The Horsehead is located about 1,600 light-years away from Earth. The area shown in this image is quite large on the sky, covering about five times the area of the full Moon. This false-color image was created by combining emission-line images taken in hydrogen-alpha (red), oxygen [OIII] (green) and sulfur [SII] (blue)"

Minimum credit line: T.A.Rector (NOAO/AURA/NSF) and Hubble Heritage Team (STScI/AURA/NASA). This image was specifically used to promote the educational aspect of the advancement into space-time by Educational Outreach. However none of the fore going organizations mentioned are meant as an endorsement for this writing.

(outreach@noao.edu)

National Optical Astronomy Observatory

Einstein, Me and the G.O.D.

(Grand Original Design)

© Copyright 2014, 2016 2nd addition

Kenneth R. Stauffer, PhD.

Einstein quotes found throughout the book are denoted by either images or *"Bold Italic print."*

Bold print is also used to interject my thought into Wikipedia etc. by paraphrasing. Paraphrasing is used in these instances for direct correlation to the issue at hand and my alternative point of view.

Published by CreatSpace

Book designed by Kenneth R. Stauffer, PhD.

Table of contents

After thought

Upon review of the manuscript, I found that figures 02A and 02B had two different spellings for hight/height. Looking up this discrepancy I noticed that 41% of the population spell it incorrectly as hight, so this eliminates any chance of blaming it on Dyslexia.

I imagine that other miscues throughout the book have also gone undetected.

American Pharoah made history, becoming the 12th Triple Crown winner. His owner, family and friends must be very proud of his accomplishments even thou they probably get razed about the spelling of his name.

Too often we are judged by what other folk's perceive what intelligence is.

Intelligence has been defined in many different ways such as in terms of one's capacity for logic, abstract thought, understanding, self-awareness, communication...

"Anyone who has never made a mistake has never tried anything new."

The glass is always half full!

Part I

Commonality of Einstein and Me

Much has been written about Einstein's life, but to me the most fascinating aspect of his being was what I learned when I was in the third grade.

The year was 1949. I was only 6 or 7 and was considered to be a slow-learner child. My mother and my teacher, at the little country school that I went to, did not think I would make it to the next grade without special attention such as remedial reading and tutoring.

My mother was on welfare and struggled just to put food on the table and clothes on our back. We had a three holler out back of the house and a Sears Roebuck catalog nearby to take care of our personal needs. A bright red painted pump supplied our drinking water and bathing, which we did in a small round galvanized tub.

Among the many chores stifling my mother, such as using a wringer washer to wash our cloths, hanging them out to dry, gathering eggs and getting milk from my grandfather's farm, she had no time or energy to help me or my two sisters and brother when it came to our studies.

I struggled through school that year and by some divine intervention I managed to make it to the third grade. It was only then, as a third grader, that I saw a glimmer of hope. Hope to compete in school and possibly get an 8th grade diploma which at that point was very much in doubt.

I was at my grandfather's house and was sitting at their kitchen table looking at the want ads of the Buffalo Evening News from the day before .I usually did this when I was visiting Grandma, when she was busy doing her chores.

I used the want ads as a means to identify words that related to things I knew, like for-sale cars, tractors, homes, and even jobs available. I was hoping that when and if I got through the 8th grade, I could work as a factory worker in a local canning plant in the next town over, Springville, about 12 miles away.

"Education is what remains after one has forgotten everything he learned in school."

But that day was a little different. I actually saw something on the editorial page that caught my eye.

It read something to the effect that a German man, Albert Einstein, was considered the smartest man in the world. I didn't understand much of the article but I asked my Gram about it that night after supper.

She told me that Einstein was not only the smartest man in the world, but that he, as a young boy had a hard time in school and that he did not even speak until he was around 3 or 4 years old. She said he used his math and science knowledge that he got from attending school. Later, when he was a patent clerk, he had his "thought experiments" published. Dr. Max Planck was probably the first one to recognize Einstein's genius!

Grandma and most others also said that his discoveries were too complicated to understand and not to worry about it.

But by this time, I was thoroughly overtaken by the knowledge that a young boy, such as Einstein, could grow up to be so smart. I wondered silently if I could ever overcome my difficulties in school and in a very small way (certainly not to the levels that Einstein reached) would be able to think the way he does. I fantasized that I became an inventor, such that my inventions would be accepted by a patent reviewer like Einstein.

I was a dreamer so it wasn't hard for me to think up ideas that I hoped could be patented, but at that age most of my day-dreams were about running away from home with only my bathing trunks on and living in the woods.

At that time, the Nabisco Company sold Shredded Wheat in boxes that had three card board divider cards to separate the different levels of wheat biscuits from each other. Printed on the cards were information about my Indian friend, Straight Arrow and his camp sight, tools, and other paraphernalia typical for an Indians life style.

I had collected most all of the series of Straight Arrow cards from Shredded Wheat boxes, so I had a good idea of how I would survive in the woods by making a shelter plus hunting, fishing, and gathering berries and apples for nourishment.

Sometimes I would fantasize that I would sneak back to the house and return my trunks, so I would be completely independent of any help from others.

These dreams where very short lived however and they mostly resulted from frustration of not having a father at home and being welfare supported.

In this way my background was not like Einstein's, but as to my learning disability, I felt a strong bond with him.

It was only when my thoughts (or Thought Experiments) were cultivated that I truly saw the parallel between his creativity and mine.

However, this was a very gradual process and I often second guessed myself as to whether I was being creative or just being a dreamer.

One example of this was around 1970-71 when I was working for the R.T. French Company as a Food Research Scientist.

I had already graduated from a two year school, SUNY at Morrisville, '63, University of Georgia, '65 and completed a tour in Vietnam, '67 in the USMC as a 0311 rifleman.

I also worked as a laboratory technician at Hooker Chemical on Grand Island N.Y., 68-69 before joining French's in '70.

At French's I was given a project to characterize the properties of mustard mucilage.

I would take the hulls of yellow mustard seed from the factory's by-product bin to my lab and mix it with water.

After it was fully hydrated I would centrifuge it to separate the insoluble cake from the viscose mucilage.

From this I would precipitate out the polysaccharide polymers by adding 2 volumes of isopropanol alcohol.

Next I would repeat this step 2 or 3 times until it was pure enough to freeze dry.

I would use this freeze dried sample to characterize the mucilage.

I developed all the viscosity and other functional tests on it during my characterization studies. Its viscosity profile showed it to be only a low to medium viscosity builder.

One important property stood out, it showed the ability to emulsify oil-in-water.

At that time the hydrocolloids were considered only as auxiliary emulsifiers by way of their thickening ability.

Thinking about this, I had a "thought" about how to test for this unusual emulsification property.

Note: this is not a "thought experiment", since my thought can be physically tested.

Since the mustard mucilage was a poor viscosity builder but a good emulsifier, I thought it might have emulsification capability due to its surface activity and its interface tension between oil and water.

At this time (1970) gums, starches and mucilage were considered to stabilize an emulsion by way of Strokes' law of thickening the aqueous phase only and not by any surface activity.

To test my theory I needed a DuNouy Tensiometer.

4

But at R.T. French they did not have a DuNouy to test my hypothesis nor did they have the money in their budget to buy one.

Some of my coworkers even thought it was a hair-brain idea since everyone knew that true emulsifiers were hydrophobic and hydrophilic.

This did not set well with me. You might say that I could be as stubborn as Einstein!

I came up with a new strategy, I would make my own simple device. It was based on a drop weight method to measure surface tension.

I went to the machine shop and asked them to make me a drop weight apparatus out of capillary glass tubing. I gave them a drawing and they said they could do it.

Finally, after about a week and a half and me persistently checking in on them, they did present me with the finished product.

I went to work immediately timing and weighing the many drops of mucilage and the pure water control.

Later that night I calculated my results and to my dismay, I was off by an amount of one decimal point.

I spent the better part of that night trying to figure out what was wrong. The surface tension of water was off by a factor by ~10 and my mucilage results didn't seem plausible.

When I showed my calculations to my coworkers they couldn't make sense of it either and chalked it off as just an inaccurate apparatus and faulty reasoning.

Now, after I found my mistake I finalized all my results and presented them to my supervisors.

I showed my boss and his boss, they were skeptical. Skeptical, yes, but uninterested?

No.

Later I learned that Einstein made similar errors when he was determining the viscosity of liquids with small spheres in suspension such as casein protein in milk.

Einstein calculations were not always correct but his concepts and imagination was "spot on".

One of his famous quote is;

"Do not worry about your difficulties in Mathematics. I can assure you mine are still greater."

I think he got this one wrong, my math pales with respect to his. However he did go on to say;

"God does not care about our mathematical difficulties. He integrates empirically."

"The only thing that interferes with my learning is my education"

They immediately ordered a DuNouy tensiometer, and ran a surface test as well as an oil and water interfacial tension test.

To my boss's amazement, and to my delight, the reading from the DuNouy was the same as my crude readings from the Jerry-rigged drop- weight glass tube experiments I concocted.

They confirmed my hypothesis that mustard mucilage was a bi- functional emulsifier by not only adding viscosity to an emulsion but also by lowering the tension between oil and water to stabilize emulsions.

Emulsions, like salad dressing and mayonnaise.

I later suggested that the hydrophobic moiety came from the cellulose part or its chemistry and also from any exposed methyl groups that could come in contact with oil in an emulsion. The hydrophilic property relates to the uronic acids of its polymers along with the many hydroxyl groups exposed to the aqueous phase.

I wrote up my results as a scientific paper and with the help of my boss's boss we published in the Journal of Food Science where I was mitigated to the third author of the paper.

Ken Stauffer Accepts Research Position in New York Import Firm

Kenneth R. Stauffer, son of Rena M. Kruse of 128 Waverly St., Springville, has accepted a position as Technical Director for the Tragacanth Import Corp. (T.I.C.), 141 E. 44th St., N. Y., N. Y. His new duties will include supervision of Research and Development, Customer Services and Quality Control for the Water Soluble gums the firm sells to food companies and other industrial users.

You might remember Ken from the Class of '61, where he was active in football, basketball, tennis and Student Council (Treasurer).

In '63, he graduated from SUNY at Morrisville and '65 received a B.S.A. in Food Science at the University of Georgia. From there, he enlisted in the U. S. Marine Corps and spent 13 months in Viet Nam as a rifleman and later as the Battalion Commander's Aid. Ken received a Presidential Unit Citation for duty with the 3rd Battalion, 9th Marines which suffered heavy casualties while operating in the DMZ.

After that, Ken taught Chemistry in North Carolina, was a Chemical Technician at Hooker Research Center and lastly, worked for the R.T. French R&D as a Food Chemist.

At French's, Ken worked on such projects as improving their worchestershire sauce, developed some dry mix products and just recently completed a 1-1/2 year study on mustard mucilase (a water soluble gum). These findings will be presented by his supervisor to the National Institute of Food Technology at their annual meeting which will be held this year in Miami.

Later, the paper should be published in the Journal of Food Science. The preceeding project which uncovered many underlying physical and chemical properties of mustard has helped Ken to gain much insight into the

My boss gave a presentation on the paper at the National Food Science Convention held in Miami, in which I even finagled my way to attend. I was not too bitter about being the last name on the paper or not being the presenter of the paper, but I did resolve the fact that I needed a change in environment if I was to find my identity.

But on the brighter side of things, I was very pleased with myself because I was able to take a raw thought, which seemed foolish to others, and have it result in a scientific paper, a scientific presentation, as well as a commercialized produce sold in the market place as a component of ground mustard bran.

Later in life I found a Canadian article about mustard on the internet that basically copied our work in their publication.

This experience was one of the first signs that made me feel creative and not just a dreamer. But I still didn't have all the answers that I was looking for about my often unorthodox way of looking at things and seeing very differently than others around me.

From R. T. French in central N.Y. to N.Y., N.Y. I found myself in a new town, a new job, a new title and perhaps a new lease on life.

I found a job in New York City as a technical director for a gum import company.

I became Tragacanth Import Corporation's Technical Director. I used this opportunity to my best advantage.

The gum import company was a very lucrative business, but it only required one technical person. I was it.

I was the "chief cook and bottle washer", so to speak. I worked in a tiny little room with workbenches put in so I could run my viscosity and other quality control tests. The room I worked in was designed to be an office so I had to use the bathroom for my water and cleaning source. I washed my glassware in a hand sink next to the commode.

Whenever I washed any glass ware, or simply washed my hands, I could look out the window above the sink and see the Chrysler building tower.

I often wondered if the legend was true about the top of the Chrysler building. The legend claimed the designer, William Van Alan, put a john at the very top of the building, just under the antenna-like towering edifice, so he could crap on NYC. If it was true, that such a john existed, I was sure that William must have used it when Walter Chrysler refused to pay his fee for his architectural work.

I learned the water-soluble gum business from the ground up. Subtle techniques, like slurring karaya gum with alcohol and simultaneously adding water, so as to get perfect hydration without "fish-eyes" in the colloidal suspension, was critical, and was part of the learning curve.

I soon mastered the practical aspect of the gum QC operation. I started to fiddle with gum blends, finding synergistic interactions with each other, which provided an increased index of viscosity at a reduced cost. Also, I made stable salad dressing emulsions with synergistic gum systems that were not conventionally used in the industry.

This "added- value" product line launched TIC into a new level of business. Instead of just importing and selling gums, they were now selling complete stabilizer systems at premium prices. Said another way, they were indirectly selling technology.

Besides salad dressing stabilizers, I developed other synergistic systems to improve or cut costs in various fields, such as paper, tobacco, batteries, tanning of hides, and paint, as well as the food industry.

After a couple of years, I worked out a deal in which I received royalties for my blends. It became so lucrative that I decided to train a technician to handle the QC and I applied to the University of Rutgers to get my Ph.D. in Food Science.

I was armed with a quality industrial research background and was also aware that I have dyslexia.

I was diagnosed by my analyst who was working with me, almost in a life and death situation for the first year of analysis.

The reason I had an analyst came quite by surprise.

I had only been at work for a couple of weeks and one of the owners called me in to look at my research notebook.

The owner was a Yale graduate and was quite worldly. He spoke several languages and was also a polished salesman.

He was disturbed to see in the notebook that I referred to the particle size of a gum as "find mesh" as opposed to "fine mesh."

I explained, the best I could, that I always had a spelling problem but it never kept me from doing good work. I cited the mustard gum research paper that I was a part of as an example.

The owner seemed to understand and told me that he had a "shrink" for a neighbor. He thought I might have dyslexia. He went on to say that Governor Rockefeller was the first one to be diagnosed with this. He went on further to tell me that it was a type of learning disability that could cause people to see and write things backward.

This was very exciting but scary.

It was exciting because, maybe for the first time, I could understand why I always had so much trouble getting good grades and why I made so many mistakes.

It was scary, because I didn't know if I really had it and scary because I didn't know if I would keep my job if I didn't.

The owner made an appointment for me to see the "head doctor." I waited anxiously all the rest of that week. I was to be tested on Friday.

Upon arriving, I was greeted by a warm middle-aged lady who introduced herself as Dr. Helen Wellington. We spoke for an hour and a half, mainly talking about my childhood. As the time kept slipping by, I kept wondering if she was going to give me a test and determine if I had dyslexia or not.

After two hours, I asked about dyslexia. She told me it was quite remarkable that I found ways to adjust to my learning ability without knowing what was going on. She further went on to say that I had other issues that I should consider working on.

She said that a lot more than what I remembered on the farm probably took place and if we could get to the bottom of it I would surely feel better. I was surprised. I thought that my war experience was what kept me from feeling good about myself.

My escapades seemed to escalate now that I was residing in the "Big Apple." The bars and night spots were plentiful and I frequented as many as possible.

Once, I got so drunk in Chinatown were I wound up sitting with a couple of hit men at a bar. I proceeded to buy them all drinks and then staggered out, only to meet up with a gang of Chinese teen thugs.

This encounter ended up in a brawl and I narrowly escaped by fending them off and jumping in a taxi. The cab driver saw the whole thing and honked his horn then waited so I could make my getaway.

Another time, I wound up sleeping in Central Park all night and was not able to recall what had transpired. The last thing I remembered was riding in a cab with a few New York Jet football players and vaguely realized that I was dumped in the park after going berserk because of alcohol and combat fatigue. My only lifeline was my analyst and she had all she could do to keep me stable. I was seeing her three times a week for several months before she was comfortable enough with my progress to cut back to twice a week.

Through all of this stormy lifestyle, I was able to manage to write a feature article for a trade magazine on gum Tragacanth and I also wrote a chapter on this same gum for a noted book company.

I did all this while without a desk in my lab or living quarters. I wrote at my laboratory work bench, on the subway going to my analyst and also on the train to Highland Falls, where I lived after being kicked out of the YMCA for sneaking a bar fly into my 9 by 12 room.

I was accepted to Rutgers and was offered an assistantship. Along with the assistantship and my GI Bill, plus the royalties I received from TIC, I was in good shape financially.

Also, the assistantship allowed me to have health insurance, which enabled me to continue my sessions with my analyst and attend college.

Before I resigned from work, and before I went to school, I got a surprising phone call from my brother. My brother explained that our father was dying of cancer in a Buffalo hospital. I had not seen my father since that day, when I was only four years old, and my dad was walking down our dirt

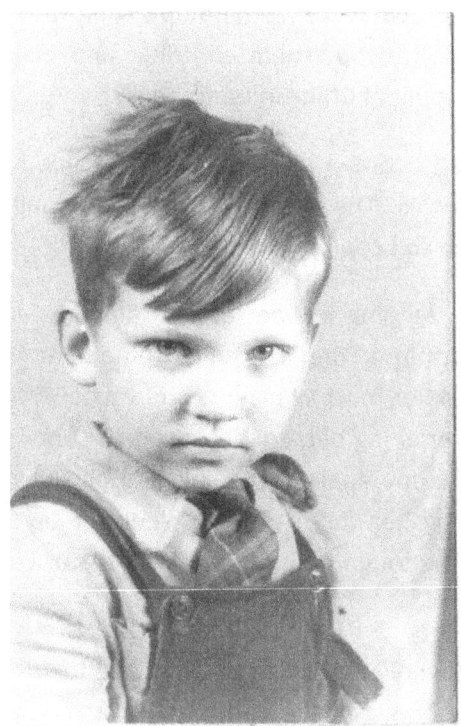

road with his suitcase.

He never came back.

I was taught that I should always be faithful to only my mother when it came to these matters. Somehow it didn't feel right if I did not get a chance to see my father with my own two eyes and judge for myself how I felt about everything. I spoke with my analyst, not to get her opinion or her approval, but just to tell her about this new situation and that I was flying out to Buffalo that night to see my father.

It really felt strange and it brought tears to my eyes when asked where I was going by someone at work and I responded, "I'm going to see my dad!"

I flew to Buffalo and was in the hospital the next day. I sat in a chair next to my father's bed. When asked who I was, I said, "I'm Ken, your second son from your first wife, Rena Mae Davis."

I found out from my brother that my dad had been married, or at least had children, with several women. He also was told that he had a dozen or so half-brothers. I sat with my father for several hours and went away with the impression that my dad was a mild man, but somehow couldn't get together with anyone to make a real commitment. I left the hospital with a lightened heart. Many of the heavy burdens that I had been carrying around for so long had been lifted.

I was ready to take on a new phase of my life—life as a graduate student at Rutgers University.

I was accepted to the Ph.D. program without having a Master's degree. This was quite unusual at Rutgers Food Science Department. Actually, it was academic since I would have to eventually pass an oral, as well as a written, test to qualify as a Ph.D. candidate.

But what it did, it helped me to keep focused on the Ph.D. requirements while accumulating credits toward the degree. For the first 18 months, I did preliminary research in preparation for my Ph.D. thesis, along with coursework to get a Master's degree.

This was a nice milestone, but I was already focused on the written Ph.D. qualifier.

I outlined my entire notes from all the classes that I took from the professors who were going to submit questions for the test.

I shared these notes with a fellow student and by laying out this strategy to him, it became so familiar to me that, when I took the test, I breezed through it with great detail and clarity. I was confident I would pass.

Also, while I was doing my Ph.D. research, I was simultaneously writing my Ph.D. thesis. I was not feeling pressured, the normal pressure a candidate might ordinarily feel, because:

Number one: I had more outside industrial research experience than most professors.

Number two: I was totally comfortable with my dyslexia analysis and I was not afraid to let any professor know that I was not to be judged by any lack of grammar efforts without the understanding that this was a flaw due to my learning disability and not because I was stupid.

Einstein did not have this luxury because dyslexia was not diagnosed yet.

Number three: I met with my analyst at least once a week, which kept me in proper perspective.

My thesis research involved the growth of an exocellular gum- producing organism that is found on trickle bed filters in sewer treatment plants. By aerobic fermentation, I was able to grow, isolate and purify the microbial gum for functionality and chemical elucidation.

The organism was unusual. It acted as if it was an ionic polymer but it did not show any uronic acid (ionic) associated with it to explain its functionality as an electrolytic polymer. Instead, its analysis was as a neutral polymer with no acid groups attached.

My big breakthrough came because I was able to "think outside of the box." Instead of looking for the uronic acids attached to the Microbial gum, as was done by others at Ohio State University and elsewhere.

I decided to look for a ketallic covalent attachment of pyruvic acid, since it was an electronegative molecule that had recently been reported to be part of another exocellular commercial gum, Xanthan gum.

I gently hydrolyzed the gum to release the pyruvic acid from the molecule and in turn made a bromine pellet and ran an Infer Red light scan to determine if the gum showed an identification-peak where the pyruvic acid carboxyl group would lay if it was found to be attached to the backbone of the gum. I was very excited when I saw that the Infer Red analysis showed that it did have a pyruvic acid attached to the gum.

I needed to repeat the experiment again to convincingly show that it was real and not an artifact but, when I repeated the test, the pyruvic acid peak was conspicuously missing. I was confused and went to bed wondering what happened. In the middle of the night, a realization came to me. In my haste to run the second conformation experiment, I overheated the embedded gum in the bromide salt so that the excess heat cleaved off the pyruvic carboxyl unit that identified it as pyruvic acid.

I went back to the lab and carefully heated another salt sample so as not to cleave the carboxyl group. With this proper heating procedure, it again showed the presence of pyruvic acid in the gum.

With this concern put to bed, I also ran a paper chromatography analysis test for another experimental type of conformation of the pyruvic acid. This test, which was run in duplicate, also confirmed my initial findings.

Now, with this new understanding that the exocellular gum of the microorganism was a polyelectrolyte, I decided to illustrate this with a pictorial representation of the organism attached to a trickle bed filter in a sewage processing plant.

I defended my Ph.D. thesis with this research along with further functional and chemical research of the microbial gum in food systems.

15

My graduate studies were complete and I was able to obtain a Masters and Ph.D. in four-and-a-half years. I put out feelers for a teaching-research position and found one at the University of Rhode Island.

I packed up all my belongings in a 1967 turquoise Mustang convertible and started on a 100-mile journey from New Brunswick, New Jersey to Kingston, Rhode Island. It was a good thing that it wasn't raining because, after I filled my car completely up with my belongings, I put my double mattress and box springs on top of the filled back seat and top of the trunk. The mattress and box spring overlapped the trunk a little bit, but the center of gravity was squarely at the middle of the trunk.

Without tying anything down I took off sometime in the afternoon, driving with one hand on the wheel and the other hand and arm holding and resting on the mattress to keep it in place. I didn't drive too fast at first because I didn't know if this packaging arrangement was going to work. But, by the time I got to the New Jersey Turnpike, I was starting to feel like I might make it. I only had 90 more miles to go.

As I sped along the three lane highway at 60 mph, I saw that the aerodynamics were in my favor. The air that rushed over the front end of the car actually helped hold down the mattress and box springs as it flowed over the top. Only on occasion did I need to adjust the mattress on the box springs while negotiating down the road. I got to the New York City tollbooth and was concerned that I may be turned away and not allowed to go through the Lincoln Tunnel. But it seemed that my money was as good as anyone else and they let me right on through. I made it through New York City without a hitch and was nearing the Connecticut tolls. Connecticut had exact change, as well as manned stations. I checked my change and saw that I could use the machines and thereby avoided the tollbooth people.

I made it as far as Gilford before I needed gas. I was tired anyway, so I pulled off the Connecticut turnpike and stopped at a closed gas station where I caught a little rest.

When the gas station opened the next day, I filled up and was off.

As I entered Rhode Island, I was starting to get a little excited about my new beginning. I stopped at the Rhode Island information center and picked up all the literature that I thought pertained to me. I got a state map and remembered telling a friend that I was going to Rhode Island and, because it was the smallest state in the union, the map was actual size.

I finally arrived at my destination, mattress and all. I had found a place to live earlier from the university listing.

I settled in and started teaching Introduction to Food Science, Biochemistry of Food and Food Processing. The reason the position was available was because one of the Food Science and Nutrition professors was on a year sabbatical and I was merely a substitute for him for one year. I made the best of this opportunity to gain experience in teaching at the college level, in hopes to land a track- tenure position elsewhere.

Near the end of the first year, I was informed that the professor on sabbatical was going to stay another year in Saudi Arabia. I was offered another year of teaching at RI.

I stayed that next year and several more after that, each time as a yearly replacement for the professor in the desert.

I graduated several Master's Degree students and one Ph.D. student during this time period. I published over 20 papers and presented as many oral research findings at various International Food Technology (IFT) Conventions throughout the US of A.

After several years teaching and doing research, I was promoted to an Associate Professor.

Along the way, I married another Ph.D. student from Rutgers and had a son with her.

To my surprise, the professor in the desert of Saudi Arabia decided to return to the university. The department couldn't absorb my salary so, after a petition of some of my students to convince the Dean to keep me, I was given my severance.

Without any university connections elsewhere, I returned to industrial research, this time working for Campbell's Soup Company as a Food Research Leader.

One of the most bizarre situations in this industrial research setting was when I was put on a project called "raw chicken to the soup can." The project was ill conceived by an entire independent group of scientists for the company, which was a satellite to it, and was operated out of an Arkansas Animal Science Campus.

The project should have been dead before it even got started because it was not legal for raw meat to be processed by the same people, or even in the same building where other foods were processed.

There would have to be an isolated raw processing line that would be used to fill soup cans that had the broth and other ingredients to make up the complete soup. However, the canning machines with the soup, minus the raw meat, would not be allowed in the same building as that of the raw meat.

Since I was a national renowned gum expert, I was put on the project to find a way to keep the raw chicken from "feathering off" when the raw meat was met with the hot broth in the soup can.

I had previously implemented a process to maintain liquid (referred to as purge- liquid that leaks after a thermal-processed meatloaf was frozen and thawed) of a turkey, and/or chicken loaf by use of gums and hydrocolloids, such as carrageenan, locust bean and starch.

This project was saving an estimated $6,000,000 a year and the gum treatment also improved the moistness and taste of the sliced poultry meat when incorporated with the TV dinner that it was used in.

I, along with my right-hand man, Joe Conte, looked at ways to utilize gums and hydrocolloids to prevent the sloughing of precipitated soluble protein (to flakes of insoluble protein) by thermal denaturation when heated in the soup can.

Joe and I saw early on that the "raw chicken to the soup can" concept was an unacceptable and illegal process as well as an impractical one. We decided to develop a process that would heat the raw chicken just after it was plucked by rapid, short time cooking to render it legally cooked, but not lose flavor. This technique also locked in the juices so the chicken would taste juicy and succulent, instead of the "sawdust" quality that currently was in the soup cans.

We did this the same general way that I improved the chicken loaves for the TV dinner. We would marinate the poultry with salts and hydrocolloids, but then, unlike the loaf process of cooking in a water bath, we used direct flame heat to cook and seal the individual poultry pieces.

This project was slated to save the parent company $25,000,000-$40,000,000/year for the foreseeable future. That is why there was so much scrutiny about who did what and who reported what and to whom.

Obviously, the Arkansas satellite team did not want the corporate research team to get credit for the execution of the project. Also, to make things worse, the corporate team did not want to let a little two-man operation design and perfect a process that represented such huge savings.

Joe and I were the only two people in the entire company that got it right, but how could we get our message across to upper management who reported to the CEO of this giant world

Corporation? Every time we tried to go up the corporate ladder, the Arkansas boneheads stiffed us. We finally decided to go to a third party vice-president (VP) who was recently promoted to product development. He previously was my VP before he was promoted, so there was a window to the CEO to lay out this vital process.

Timing was getting very tight. Joe and I were just at a huge inter-departmental meeting where the Arkansas research team was outlining the "raw chicken to the soup can." They totally brushed off the legal aspects of the process saying that they would address that situation later.

Joe and I got through to the product development VP by way of a mutual project leader who worked under him. He believed in what Joe and I were trying to promote.

Joe and I set up the flamed chicken stations and wrote reports about this research approach for the project.

Funny though, we were not allowed to go to this closed meeting that was held in the late evening. Only invited management and corporate representatives were there.

The next day, Joe and I went to work, anxious to hear the outcome of the meeting. It took a while but finally we caught up with our mutual project leader at the end of the day. Everyone else on our floor was gone and he made sure no one was around to hear him. The project became very political and he didn't want to be associated with us for fear that he might be looked at as a non-team player. He told us that the project "raw chicken to the soup can" had been dissolved and a new project was proposed, "Project Chicken."

This was all he could say except that he thought that we should just accept whatever came down and be team players for the new project.

For the next few months, Joe and I were on board, following the new plan that ironically looked exactly like the one we recommended.

Over the next year, the project became a reality and one by one the soup plants all over the USA and Canada were using this Technology. The project was nominated for and won "The Project for Outstanding Achievement" award. It was the second highest award that the company gave out, only the coveted "Crystal Award" was higher, and it came with a stipend of $10,000 dollars per team or individual.

The next year, however, "Project Chicken" was nominated for this award.

To the surprise of the corporate research group, they saw that Joe and I were not mentioned on the slate of team members. The Arkansas director failed to put our names on the team, sighting that it was a different team that implemented the project into production. The corporate team would have no part of this. The president of research made sure that we were included in this nomination.

At this point, Joel and I didn't care whether or not we were included in this nomination. For that matter, we didn't care if the project won at all. Actually, the project came in second to a marketing project that was based on brand identification, where chicken soup was promoted by putting more chicken in the soup and selling it for the same price.

I had never been exposed to such cut-throat and backstabbing before. Yes, there were maneuvers at URI and the import gum corporation to gain advantages over competitive associates, but never before did   I see such "no holds barred slug fest" that this project nurtured.

Reflecting back, this experience seems (in a small way) similar to what Einstein went through when he was nominated for the Nobel Prize. Below is a parcel review of *Aant Elzinga's book ...Einstein's Nobel Prize: A Glimpse Behind closed Doors*

Sagamore Beach MA: Science History Publications, /USA 2006, ISBN: 0-88135-283-7published in shortened form in Brit. J. Hist.Sci. 41.i (no.148) 148-149, March 2008

"Each of Albert Einstein's three ground-breaking papers of 1905 is nowadays considered to have merited the Nobel physics prize. He was nominated from 1910 onwards by increasing numbers of leading physicists. Only in 1922 was he awarded the deferred 1921 prize and on very narrow grounds, namely for the law of the photoelectric effect. How did this happen? Elzinga's book based on the Nobel archives illuminates the tortuous path taken - from a narrow interpretation of Alfred Nobel's terms to antagonism on the four-man physics committee and a fudged compromise when the international credibility of the prize was at stake. As Nobel-laureate status is for many scientists the highest aspiration, the awarding system is significant in direction setting and its functioning has importance.

The awards influenced the key developments of 20th century science, for good or ill. Einstein could have been awarded the prize for his contribution on Brownian motion, revealing the molecular structure of matter and forming the basis for statistical physics. His award could have been for contributing to quantum physics, the overarching theory of the subsequent century. Or the award could have been for the special and general theories of relativity. But it was specified just for an experimental 'law' relating the frequency of ultra-violet light to the energies of the electrons it ejects from metals. The novelty lay in light interacting with solids on the atomic scale and in Planck's quantum constant 'h' featuring in the law.

Elzinga makes clear the Nobel committee would not reward creative science, but only 'discovery or invention'. It definitely did not reward 'metaphysics and speculation', a stance relaxed in recent decades. Though Einstein's novel concept of duality (spelled out in 1909) went unmentioned in the citation, his prize is frequently attributed to his role in originating quantum theory. The 1982 biographer, Abraham Pais (Subtle is the Lord. The science and

life of Albert Einstein, Oxford 1982) stated, indeed, that the award was rightly given for quantum physics (despite most nominators arguing for relativity) and lauded the "judgement of a highly responsible rather conservative body of great prestige.... the story has neither heroes nor culprits" Elzinga's judgement is different and his analysis pulls no punches, showing a verdict cobbled together, amidst personal biases, for a narrow experimental law. Chance had intervened with the unexpected death of the committee chairman. The hero was a new committee member (C W Oseen) who saw the others were set against relativity, so argued for the 'law' being fundamental and underpinning Bohr's atom model. He successfully argued for a package using the deferred 1921 prize- one to Einstein and the second to Niels Bohr. But such was the animus against relativity theory that at the Swedish Academy's plenary meeting, the astonishing reservation was added to Einstein's certificate: "independent of the value that (after eventual confirmation) may be credited to the relativity and gravitation theory". This reflects the refusal throughout the decade to award a prize for relativity, revealing inadequacies and prejudices of committee members *that Elzinga exposes in detail. The records for 1910 and 1912 show they should have awarded the prize for the special relativity of the 1905 paper, which synthesized nth century electromagnetism and the finite speed of light, while predicting increasing mass of an electron at relativistic speeds. By 1914-15 experimental tests confirming this.*

The review was by **MAX K WALLIS, Cardiff University and TREVOR W MARSHALL, Manchester University**

Too many, and to Einstein himself, this convoluted treatment of distorted half-truth felt like a slap in the face.

Eddington's' eclipse experiment proved Einstein's theories, but skeptics argued that the trouble was that Eddington's' observations had not been perfect since he had discarded data he considered poor from his observations. In Jeffrey Crelinsten's' Einstein's Jury, this smelled like cooking the books to Einstein's advantage.

In reality Eddington was just employing good scientific practice.

The final result of this debacle is that Einstein did not attend his Noble prize ceremony. Despite being informed that he was to receive the prize, he chose to continue with his lecture tour in Japan. This was probably because he was disgruntled about the politics of the award.

I too was also very disgruntled. I looked forward to my early retirement at 55. I felt that the research was fun and challenging, but the bullshit that went along with it was devastating. The biggest part of the problem was that the bullshit time was taking over the quality research time. I was burnt out and wanted out.

This did not sit well with my home life and within a year I had left the company and divorced my wife. I truly hit bottom and needed to find myself.

Time went by and I was able to gather my thoughts about what was going on in my life.

I came to the conclusion that, as far as my experiences went in the corporate world, I was partly responsible as the directors, managers and VP's were with respect to how all the different teams became fractioned into their own little power games. I now felt, if I would have listened a little closer to my heart instead of my emotional logic, I might have been able to defuse the situation a bit so that the fractured teams might not have resorted to such huge power plays. When the physical world relies only on the five senses and stays completely in the animal world of "survival of the fittest," then conflicts like this are inevitable. I also thought about my dissolved marriage. I knew in my heart my wife and I had different ideas about life and I felt that separating and getting a divorce was the inevitable.

When we married, we were two lonely people who found each other out of common need. It turned out that this common need was not a strong enough bond to keep us together.

However, it was important to nurture and cultivate my son's life in the best possible way. Although this was difficult at times because over the years I was kept isolated from him by his mother.

I imagine that Einstein and his first wife had a heavy heart when they had to deal with the loss of their first born.

Einstein used his Nobel Prize money to set his first wife up financially after his divorce, were as, I gave my first wife the first million dollars of our saving plus the house and car. Most of our acquired money came from my trade secret formulas that I developed in my spare time.

War and Politics

Einstein was considered a pacifist when it came to military conflict, however he was also one of the twentieth-century's most outspoken political activists.

Einstein on Politics: Hedonism, War, Peace, and the Bomb written by **David E. Rowe & Robert Schulmann and promoted by Princeton University Press stated:**

"deeply engaged with the events of his tumultuous times, from the two world wars and the Holocaust, to the atomic bomb and the Cold War, to the effort to establish a Jewish homeland, Einstein was a remarkably prolific political writer, someone who took courageous and often unpopular stands against nationalism, militarism, anti- Semitism, racism, and McCarthyism. In Einstein on Politics, leading Einstein scholars David Rowe and Robert Schulmann gather Einstein's most important public and private political writings and put them into historical context. The book reveals a little-known Einstein--not the ineffectual and naïve idealist of popular imagination, but a principled, shrewd pragmatist whose stands on political issues reflected the depth of his humanity.

Nothing encapsulates Einstein's profound involvement in twentieth- century politics like the atomic bomb. Here we read the former militant pacifist's 1939 letter to President Franklin D. Roosevelt warning that Germany might try to develop an atomic bomb. But the book also documents how Einstein tried to explain this action to Japanese pacifists after the United States used atomic weapons to destroy Hiroshima and Nagasaki, events that spurred Einstein to call for international control of nuclear technology."

I on the other hand consider myself as an independent middle of the road conservative having strong nationalistic ties to my country.

When I was 23 years old and just about to graduate from the University of Georgia, I had the philosophy: "that no one should be drafted unless there was a war going on."

Little did I know that my convictions would be tested that spring (1965) with the announcement of the increased USA involvement in Viet Nam by sending troops over there to stop the domino effect of communism spilling over into neighboring countries in South East Asia?

It took a while to sink in but I knew what I had to do. Sometimes doing nothing is a way to make a decision. I left school after learning I had passed all of my courses. I did not stick around school for the actual graduation ceremony because I had no family or loved ones to share it with. I left for home with a heavy heart and depression. Depressed, not because I knew I must honor the USA draft system, but depressed because I was in limbo and couldn't make a hard and fast decision about the military.

All that summer and into the winter, I went to local bars and drank myself to sleep. I ran with my drinking buddies and would get into bar brawls on a regular basis. One night, I wound up in the same jail where my father was kept in for two years for leaving my mother.

I spent the night there. My mom had to come down with my uncle (step father) and bail me out.

The next night, I was back at it, drinking and staying out all night. My mother finally kicked me out of the house so I stayed with one of my drinking buddies. I was pathetic, to even think of me having a relationship with a woman was ludicrous. I couldn't get out of my own way, unless it came to a good bar brawl, and then I would fight with rage. All the pain of my childhood years, of which a lot was totally blocked out, and the loss of my one and only true love just seemed to drive me into these drunken rages. My sister once said that I acted like my alcoholic uncle "Dehart". Finally, I woke up one morning (afternoon) and realized that I was going nowhere fast.

I had been waiting to be drafted but the letter never seemed to come.

My mother had let me back in the house but I knew I needed to get on with my life.

I went to the local draft board that day and volunteered for the draft. I was called up the next week and took my physical.

I was way out of shape. I had a beer belly and was out of breath with just a little exertion. But my vitals were all good and I passed my physical. I got a letter the next week and was told to report for duty.

The first thing the sergeants told the recruits was that "one out of every seven of you is going to be drafted into the Marine Corps."

I knew immediately that I was about to become a Marine Corp boot, just like Sharon's father was in WWII. I laughed feverishly because I knew that this might well be the last laugh I would have for a long time. I was right.

I hit Paris Island in January of '66 and was considered a fat body since I had put on all that weight drinking all summer and fall and, of course, through college. I and the other "fat bodies" would always have to be last in the chow line and sit at the "fat body table." On more than one occasion, the drill instructor would herd the entire line through the serving station in less than two minutes. This caused the servers to fling food on the stainless steel trays as best they could. When I would turn from the servers for the regimented tray inspection of the fat bodied, the drill instructor would look at my tray and say, "You just ate." I would have to dump my tray and wait for the next meal for sustenance.

One week later my platoon drew rifle range duty. We spent eight to twelve hours a day pulling targets for another platoon who was on the rifle range to qualify with the M-14 rifle. They could qualify as an expert, rifleman or marksmen and receive the appropriate badge to wear on their uniforms. The lowest level (marksmen) score needed was 190 points.

I learned the secret language down in the target pits really fast. Once in a while, a target puller would report a controversial bullet mark as a positive hit if he thought it would help the shooter qualify. I didn't have any problem with this and I would always give the shooter the benefit of the doubt if any shot marking was close.

Actually, the time I spent pulling targets was the most relaxing at Paris Island because I did not have anyone standing over me, criticizing me.

The next week, my platoon was on the rifle range to qualify. I was actually quite good with a rifle. When I was a kid back on the farm, I would take my BB gun to the neighbor's barn, which had an abundant amount of pigeons. The first pigeon I ever killed, I used up an entire cylinder of BB's to kill it.

After killing it, I dressed it out and froze it for a pigeon dinner. As time went by, I got my wind-age and flight projection down so that I could actually drop a flying pigeon anywhere in the haymow of the barn. I felt that the pigeon was a part of the food chain and, by shooting it, I not only eliminated a farm menace, but I also supplied food for my table.

Now, I was in Paris Island and was preparing to qualify on the famous Marine Corp rifle range. Qualification day came and I was ready. Unfortunately, the weather was foul. It was the worst windy rainstorm that February since I hit the island. The drill instructor had told us that if anyone went UNK (didn't qualify) that there would be "Hell to Pay".

The wind was so strong and it was raining so hard that when I was shooting in the upright position with rapid fire, I only hit three out of 10 targets and none of them were bull's eyes. I had my scorecard marked and, after several rounds at 200 meters, 300 meters and 500 meters, I was scored at 180 points. I needed 10 points or 10 straight bull's eyes just to qualify as a marksmen with my last 10 shots in the prone position at 500 meters.

After scoring my card, the drill instructor scolded me and told me that I was sure to go UNK.

I got down in the prone position and knew what I had to do. I had to take every shot, one at a time, and not worry about any of the rest of the shots to follow. I double checked and readjusted my sight to the wind-age for the 500 meter distance. I also needed to adhere to the BRASS system of military rifle protocol: Breathe, Relax, Aim, Safety check and Squeeze. Squeezing the trigger eliminates any jerking effect and the piece (rifle) will go off by itself. After I got in the prone position, I took the charger out of the charger guide. I let the bolt go home, took a big sigh and totally relaxed. I aimed down the barrel and released the safety and slowly squeezed the trigger.

The piece went off and I looked down at the target pit and saw the bull's eye flag go up.

I knew that all my wind-age dope and techniques were on. Also, by being in the prone position, I knew that this gave me the stability to possibly qualify.

I repeated this technique and got two through nine targets without any distractions. Every time I squeezed the trigger, I took a long breath to see if I was still in the running for a medal.

I laid on the ground and knew that if I could find my way to just keep my composure to make one more bull's eye, then I could keep from being an UNK and being ridiculed by the rest of the recruits. A million thoughts went through my head, but suddenly I started to have a calming feeling. My thoughts turned to infinity and the finite of infinity. I thought of the big picture and tried to place what I was doing with respect to the operations of the universe and Eternity.

I focused on where I fit in the bell curve of humanity. With this, I revived my composure and went back to the prone position.

I relied on the BRASS technique for the last time and after I squeezed the trigger, I just laid still on the ground for a second or two before I looked up. These few moments seemed much longer because I took this time to check out my own feelings. I admitted to myself that this was a very important event in my life and that, even if this was not a universe changing experience, it was important to me and that was very important to my universal self and Soul.

I looked up and was gratified. I knew that there can be certain things that can only be achieved in life and I had just experienced one.

I went down to the drill instructor to get my total score and saw the 1900. The drill instructor glanced at the score and told me that I was very lucky. I looked the instructor straight in the eye and told him, "It wasn't luck. It was skill and dedication."

The instructor looked stunned. I picked up my rifle and walked away without incident.

For the next few days, the UNK's had to circle the platoon while the rest of us were marching wherever we went. Only 38 % of the Troops qualified so there were more recruits circling the platoon than were actually marching. It was quite comical watching the UNK's dash around the troop with all their field gear and their rifles held at the ready.

The rest of boot camp came and went without incident and I was set for graduation. As usual, no one was there to celebrate my success.

I was shipped out the next day for Camp Lejeune, N.C. for field training as a rifleman. Camp Lejeune was very hard, but it solidified the final physical and mental conversion of me into a combat ready Marine.

There was one feisty Marine sergeant who stood about five foot tall with elevated boots, and was obnoxious in his demeanor. One day, just before we left for home and then on to Viet Nam, he got on a wooden crate and proceeded to lecture about the war. He said, "I personally talked to the Commandant. He told me some of you will get to go to Viet Nam TWICE." He further went on to say, "Look to your right. Look to your left. One of you three will be killed or wounded in Viet Nam."

Actually, he was right, but to me, he seemed like a little blowhard and his only job was to play mind games with the new Marines.

The last day at Camp Lejeune was an easy one, after my seven-mile forced march in the morning. I was able to stay in the barracks until I was called to get airfare to go home and also money to get to Camp Pendleton. I spent a 20-day leave in Springville, N.Y. and had a five day travel allowance to get to Pendleton, California.

During my 20-day leave in Springville, I came across Barb Mac Hose, Sharon's younger sister, and, to my surprise, she told me that their mom had a baby girl. Barb invited me to come to the house and see her little sister, Amy. I was delighted. I had not seen Barb and Sharon's parents in a long time. Upon arriving, I was a little nervous, since I was not sure how I would be received.

The tension broke quickly when Mac greeted me with a beer and a warm welcome. After all, I was a Marine now, just like Mac, so we had a lot to talk about.

Amy was a little toe-head, just like her dad Mac, two of the three sisters, Sharon and Barb, and her brother Scott. Janice was a brunette just like her mother.

Her brother Scott, being the only boy, was a chip off the old man's block. He was witty, resourceful and athletic while at the same time was kind, polite and mischievous. Not bad for a kid of only 12. The entire family was very attractive but at the same time very humble. This was one of the family traits that made me want to bond with them but of course I knew in reality that my interactions with the family were limited.

I spent the afternoon with the family and ended my stay by joining them for supper. Barb drove me back to my house in her Barracuda. Barb and I were always good friends and she looked up to me as a big brother in school. As we drove, we spoke about Sharon. Barb told me that Sharon was doing well and that she was caring for her little baby boy, Timmy. Mother and daughter would now be rearing two kids. Amy was Timmy's aunt.

When I got home, I experienced a rush of excitement from being with the Mac Hose family. At the same time I had a big hollowness in my stomach thinking about Sharon and how much I missed her. I went out that night and drowned my sorrows in scotch on the rocks with Colt 45 as a chaser.

"No, this trick won't work...How on earth are you ever going to explain in terms of chemistry and physics so important a biological phenomenon as first love?"

When I woke up the next day, it was 2:00 in the afternoon. My mouth felt like someone just pissed in it and my head was pounding as if a herd of horses and a small pony had just trampled on my skull! As I had done so many times before in college, I partook in the dog that bit me the night before. I sipped beers all afternoon and into the night, only to repeat the process until the end of my 20 days of leave.

31

I sobered up enough to catch a plane to Charlotte, NC on my way to California and Camp Pendleton. Puddle jumping my way across the states, the aircraft arrived at Sheep's Chad, Camp Pendleton. I was assigned to the 27th Marines, which were newly forming and training as a complete combat unit to be a fully intact regiment reinforcement for Viet Nam. Most of the troops had been there for a few weeks and had formed cliques already, so I didn't exactly blend in. After I was only there a few days, I was caught scratching my balls in early morning chow formation. I was sent to mess duty for 30 days. This was a 14-hour a day assignment cleaning up, serving and helping in the galley. The rest of the regiment was left in their barracks to lounge around or sign out to go to the PX (Marine Corp superstore) until such time when they could start training.

I put in three solid days of mess duty. I was tired, sweaty and upset because I was the only one in my barrack who had work duty. I laid in my rack that night thinking about what I could do. When I went to the mess hall the next day, I saw that the steam jacketed kettle was scored in the pouring spout and told the PFC cook that this was a high risk source for contamination. The PFC told the sergeant and the sergeant told the Top (First gunnery sergeant) and the Top summoned me to his office.

Top wanted to know what all the fuss was about. I explained the potential danger and also explained that I had a B.S. in Food Science from the University of Georgia. I also asked if I could possibly go to cooking school so I could use my knowledge more effectively. Top told me to get out of his office and get back to work. The Marine Corp. was just fine without my input. I was down.... but not out. I would have to take more drastic measures to rectify my injustice.

The next morning, I got up an hour early and went to the infirmary. I told the medic on duty that I needed to see the nurse ASAP because I would be AWOL at the mess hall if I didn't get back in time. The medic told me to go report to the mess hall and get a pass and then return. I said that if I did that I might contaminate the food, since the sergeant wasn't too concerned about personal matters. The medic was annoyed with this

32

whole situation and asked what in the hell was wrong with me and why I thought I knew more than the sergeant in charge in the mess hall.

I responded, "I've got VD and should not be around food, and furthermore, I've got a degree in Food Science and I know what I'm talking about!"

With that, the medic sent me to the doctor and the doctor took a culture swab, issued me a pass and sent me back to my barracks. For the next 10 days, I was allowed to lounge around the barracks. The troops, however, had to finally go out in the field to start training since they had filled their ranks to train as a unit. One morning, the troops were going to be inspected by the captain and were told to "fall out." The barracks sergeant told me to hang back. He was afraid I might fuck things up. I thought that this was quite a turnaround from just being a "whipping boy stooge."

The next day, Top summoned me to his office. He told me that I was going to school. Top blurted, "It's OJT, on the job training, and I'm sending you to Viet Nam for your schooling."

I looked Top straight in the eyes and said with my poker face, "Thank you, I'd much rather be there than here in garrison, just playing cowboys and Indians."

With that said and done, Top gave me my marching papers and the next day I was in the big bird flying to Alaska for an overnight layover and then on to Da Nang.

I arrived on September 4th 1966 for my 400-day tour of duty. I had just crossed the International Date Line, so I was already down to only 398 days left. The next day I was sent 30 miles inland to the area called An Hoa, which was not too far from a place referred to as the Rock Pile, where some memorable battles took place.

When I and three other replacements arrived at the campsite, we were greeted by a couple of young privates. The first thing they asked was if we (the new replacements) had all are gear for the field. Having learned that I didn't have a smoke launcher and a canteen, the two privates looked at each other quickly and mumbled some garble to each other. Nodding their

33

heads in agreement, they told us new replacements to come with them to a different tent were additional gear was stored. Here I was given my missing equipment.

The next day, I was on a landmine perimeter sweep and I saw a hole in the trail. One of the privates told me that last week a Marine was killed there because he stepped on a landmine.

Further down the trail, I smelled a familiar but very obnoxious smell.

When I was back on the farm rounding up cows with Tippy (my dog), I would, on occasion, be exposed to this awful smell. As my squad proceeded down the trail, the odor became more and more intense. The column of troops turned the corner and there it was, a dead Viet Nam male. The troops called him a "gook".

They thought that I would be spooked, but because I had been reconstructed mentally during training in Paris Island and Camp Lejeune, my glazed eyes told the whole story and I just shrugged the day's trauma off.

Only much later did I get a proper grasp of the day?

That night, I had watch duty and the same private who got me my gear upon our arrival escorted me to my post.

It was there that he dropped a bombshell. He told me that the tent he got the gear from was "the dead man's tent," because they kept all the gear from the wounded and dead there before sending any personal items home. He went on to say that the smoke canister was from the recently diseased PFC who had been just killed along the trail.

This did stir my innards, right down to my very Soul.

As I sat on watch for the very first time in Viet Nam and for many times after, I would think about the big picture and wonder about my being and my Soul. I wondered if my physical being was inside my Soul or if my Soul was inside me! I wondered if a Soul could be found in a body. I reasoned that if a person had an out of body experience, then the Soul must be in the body ordinarily.

I felt that animals did not have Souls the same way as humans do, mainly because animals are part of the food chain below humans. If animals had Souls, then so would insects, microbes, trees and all plants. They all share similar DNA anomalies. On the other hand, Infinity is Infinity, so anything is possible. (Note: Souls are not a component of the universe, Souls utilize the universe to make decisions and thus infinity of the Soul is one of the few topics in this book where infinity applies!) This will become clearer in Part II of this book.

After much thought to this dilemma, I rationalized that I would error on the side of human experience and consider an animal and all other living things to be living without indigenous Souls.

I also went on to think that all (other than humans) living things could be carriers for universal Souls that have not yet reincarnated and may choose to slip in or out of DNA. Such Souls might be relatives, wrongly killed humans etc. Such DNA candidates for universal Souls to associate with might be butterflies, birds, ladybugs, cats, dogs, Venus flytraps, ameba, E. coli etc.

I was really meditating intensely that first night sitting on a sandbag in the middle of nowhere in north central Viet Nam. Suddenly, I heard noises and saw that my relief was coming.

I was assigned to my unit as the third rifleman in the third Fire Company, third squad, and third platoon of India Company 3rd battalion 9th Marines. My new routine at An Hoa (an outpost of Da Nang) was to go on mine sweep patrols during the day and outer perimeter ambush sites during the night.

My fire team consisted of only three troops instead of the usual four, since they were short on men to fill all the ranks. I was one of two riflemen and a rifleman leader. The other rifleman was the point (first out) and I was the tail (last in a column of three). My fire team lacked a machine gunner.

35

When we went out on patrols or ambush sites, we would go with a squad (three fire teams), which meant that about 10 troops would head out since the other teams were missing men too.

This routine lasted for exactly four days because on the fifth day after arriving in Viet Nam, the 9th Marines were on the march. It was called Operation Prairie II and the entire division was going to participate. All four companies, Kilo, Lima, Mike and India, of the 9th Marines were to assemble for a major push near Da Nang and the Rock Pile to flush out the Viet Con (VC).

India Company was selected from the four companies to be the point battalion to sweep through the jungle and push the enemy toward the other three companies positioned to block and or cut off any VC trying to escape.

The third platoon was selected to be first platoon out. The third squad was selected to be the first squad out.

The third rifle company was selected to be the first rifle team out.

I was promoted to the point and would be the first Marine out to lead India Company into the jungle on a major battle in Operation Prairie II, and I had only been in the country for about a week.

After going through a few villages and stopping long enough for the officers to question the local woman and children, I was instructed to follow a trail heading into the thick of the jungle. I couldn't help but to think of the cow paths back in West Valley and how similar they were to this, except for the six foot deep by two foot wide ditch along the side of the trail.

I and the entire India Company behind me were moving along the trail, getting closer and closer to the tree line. When I got to the tree line and started to penetrate into the darker thicker part, I suddenly saw in the six foot trench running alongside the trail, a couple of VC's with bamboo green leaf branches. They were walking away from us so they didn't see us on the trail. I automatically squatted down and looked back and, with eyes bigger than silver dollars while rapidly pointing ahead, I lip synced, "Gook,

Gooook, Gooooooooooook." The lieutenant finally got the message and brought up a machine gun team.

They shot into the area where I saw the gook and an intense firefight broke out. I was right next to the machine gun as we all fired into the area. When the dust settled and the air cleared, we found that 12 VC were supporting an AK-47 type machine gun that sat on a Russian tripod and seven of them we killed.

My ears were ringing from the machine gun and, only years later, I found out that my ear drum had broken and had healed over, but my hearing became impaired.

After this fire fight India Company quickly moved out and we that were stationed to block the VC's exit. As I passed their location, I did see two Marines fatally wounded, lying side by side, which were covered-up by their comrades.

We were moving toward an open rice paddy. I could see an Ontose in the paddy. The captain gave the order and the Ontose, along with me and the rest of the 3rd platoon, moved out. We got into the rice paddy, about the length of a half of a football field, when all hell broke out.

I dropped to a prone position and started shooting into the rice paddy. But then I heard shooting behind me and someone shouted out that we were surrounded. I spent both of my magazines and needed to reload. I looked around and saw that my rifle team leader had gone back to the tree line and was on the radio and in communication with someone. I heard one of the Marine squad leaders screaming in the rice paddy. He was hit pretty bad and needed attention, but no one could advance due to the heavy firepower.

I felt a sharp hot twitch in my right lower leg but it didn't seem to be a big deal.

Because I knew that I was surrounded and that my team leader was not there to give orders, I decided to make a break to the tree line and reload my magazines. At about the same time, the Ontose operator was shot along with the machine gun gunner.

The lieutenant of the second platoon (he was a school teacher who was a reserve and volunteered for combat duty when the war broke out) ran out and jumped onto the Ontose and started shooting the machine gun on the vehicle. He was shot in about 20 seconds and fell to the ground behind the vehicle. Someone hollered out, "Does anyone know how to shoot a machine gun?

I never had, but I knew the Marine behind me was a machine-gun operator. I looked back and the Marine made it known that he was not about to operate the gun.

Just about this time, I could hear airplanes overhead and felt good that air support had arrived. I looked up at the B-52's and saw the bombs tumbling out of the bellies of their bomb compartments. I listened for the explosions, but they never came. I thought that the bombs were all duds and said to myself, "just my luck".

But the bombing silenced the gunfire and the battlefield fell quiet.

That night, the troops were told to stay in the tree line and get some sleep. Sometime in the night, a Marine from the second platoon came over to me and told me there was a dead gook on the ground that the second lieutenant shot out of a tree just before he jumped onto the Ontose. The gook was a female and was dropping grenades on us as we negotiated the thick terrain of the jungle. of the troops when we went by.

Now she just lay on the jungle floor, dead as could be.

When the Marine realized the gook was a female, he opened up her shirt and fondled her breast. He asked me if I wanted to join him, but I told him, in no uncertain terms, that no matter what his religious beliefs were, I was certain that this was not something that he should do.

The Marine got the message and left her alone. I slept the rest of the night guarding the corpse, just in case anyone else got any ideas.

38

I knew that all the fallen warriors, male or female, friend or enemy, would all have Souls moving on to their next experience, and I knew that I would meet up with all of them in due time.

I knew nothing about karma at this time in my life, but this line of thinking just seemed logical.

The next day we were off and sweeping again. I checked out my leg and saw that I had been nicked by some of the scrap metal when I was pinned down in the rice paddy. It didn't seem serious so I just let it go.

During the sweep, I asked my lieutenant what the story was with the bombs not exploding. The lieutenant took one look at me and said, "You idiot, those bombs were napalm bombs and they don't explode, they just burn".

I did feel like a complete idiot. I had hoped that maybe the lieutenant would acknowledge that I helped save lives because I saw the VC in the trench and thus foiled an ambush the VC were setting up. Surely, many more lives would have been lost if I didn't see them. But the lieutenant would have none of that. He had an inherent dislike for me, probably in part because he knew that I had a BS degree and he was just a converted second lieutenant from a sergeant made possible by a special program to create lieutenants to fill the ranks.

Actually, I was not at all condescending about the lieutenant's background. I knew how hard it was to rise from the ranks without papers.

I found peace with myself when I realized I just saved my own life and it felt really good. As far as saving any others, it was not something I needed to think about. During the sweep, my platoon caught up with a sister company that had just trapped a VC in an open field and killed him. His body was still on the ground where it was cut down. I was summoned to the battlefront by my riflemen team leader and was told to report to the other sister platoon leader who had just been in the firefight. I went up and they told me that my job was to search the gook and that I could keep whatever was on him. This seemed very strange, but I figured, what the hell? If I get shot or if the body was booby-trapped, I might get to go home early.

I went out into the open field and searched the bullet-ridden body. I found some American change and an old brass tube shaped lighter. It looked like a lipstick holder, but I knew what it was because I had seen these old lighters when I was a kid on the farm.

I motioned to the platoon leader that it was clear and he and others came out. I showed him what I found and they sent me back to my fire team. The sweep was over and I read in the next issue of the Stars and Strips that the sweep resulted in 73 VC killed and only a couple of Marines were lost. If this was true, then I had seen the two Marines along the trail. It was anyone's guess if the 73 VC's was accurate. The body count was never explained in the article.

After the big 9th Marine sweep of Prairie II, I was back at the An Hoa camp doing perimeter landmine sweeps and nightly ambush setting.

One night, my fire team was held back to protect the perimeter of the camp while another fire team went out. Sometime in the middle of the night, a flare was tripped off outside the barbed wire. The barbwire was the camp's last defense. The lieutenant summoned us troops to the sandbag wall and ordered us to shoot into the dark of the night where the flare was tripped. I gathered up my gear and started shooting into the jungle when one of the seasoned privates crept by and told me and the others to just shoot in the air because he knew that the fire team that went out that night was sitting out there.

I was at a loss. I wanted to protect the perimeter but at the same time, I felt that the private must know what was going on. Between the lieutenant barking orders and the private giving hand signals, I decided to err on the side of life sparing. I shot into the air and the lieutenant was never the wiser.

I learned the unwritten law. The ambush sites that the company drew up were not the actual routes that we troops would necessarily follow.

More often than not, we would just go out far enough to be hidden from the camp, but not out to where the actual ambush site was drawn up.

The company Top, a gunnery sergeant, saw my file and saw that I was a college graduate. He considered the possibility to use me as a radio man for Captain Getlin, who was the commanding officer of India Company along with one of his second lieutenants, Lt. Bobo. I knew of Lt. Bobo because he was from Niagara Falls, which was not too far from where I grew up in West Valley and Springville.

After some thought, Top figured it might be better to bring me into the command headquarters group to help out with the night patrols and bunker duty. I was informed, by my captain the next morning that I had camp perimeter watch duty. I stood my watch and thought about this break I was about to get. It was like being spared from death again.

Off in the distance. I saw streams of tracers coming out of the gun ships (fighter choppers referred to as "Puff the Magic Dragons") in a fierce air support for some hostile enemy nests of VC (Viet Con) or NVR (North Viet Nam Regulars). I knew that for every glowing red tracer I saw, there were 25 regular ammo projectiles in between the tracers. Puff the Magic Dragon could cover an entire football field in a matter of minutes, or so I was told.

The next morning I was disappointed, but not entirely surprised, to learn that the company was packing up and moving out to support the fighting I had seen the previous night. My appointment was canceled with Top and I could only hope that I would make it through this battle to interview with the gunny.

The officers and sergeants were in disarray. They told everyone to assemble ASAP without any preparation instructions. The only thing I could think of was that I was not going to get caught in a firefight without enough ammo again. I stuffed my backpack with all the shells I could muster. My backpack felt like it weighed more than I did.

With all the troops in formation, we headed out. We marched all day and I only had one can of lima beans in my pack, the rest was ammo. Night conditions were setting in and the officers kept stopping and reading their maps. They turned us around and headed us back.

They marched for about a half an hour and stopped again. We turned around and marched back the other way.

This went on for what seemed to be most of the night before they told us to make camp alongside the road.

It seemed like we had just gotten to sleep when we were awakened and told to move out. After a few hours, we came to a clearing where troop transport choppers were hovering overhead and landed to whisk us away. I was never told, but I thought that moving by air was a back-up plan because the unit got completely lost and the people in charge needed us to get wherever we were going ASAP. It had been over 24 hours and I only had eaten my one can of cold lima beans.

With fixed bayonets and magazines loaded, we entered the choppers and flew out. We were in the rugged hilly part of Viet Nam. I figured it was where I saw Puff the Magic Dragon shooting tracers in the night fighting before we left.

After about a half an hour flight, we hit the landing area and the back hatch opened up. All of the troops rushed out under fire. Luckily, no one was hurt. We took a position in a small hill near the landing spot. As we dug in, I could see the battle unfold all around me just as though I was in a movie. I could see the actual barrel flames from the anti-aircraft guns of the enemy. They were shooting at the American planes and choppers in the air. This went on all night. We didn't get any combat on the hill. We were lucky.

The next day, we saddled up, still with no chow, and moved down to a valley. It was just the 3rd platoon. The other platoons went in a different direction. Four hours later, the lieutenant told us troops that we would take the hill to the left of us.

When we got there, one of the fire teams started to climb the hill. They didn't make it more than a few feet before they told the lieutenant that the underbrush was so dense they couldn't make it up the hill.

The lieutenant checked it out and agreed. Meanwhile, I went down a little farther on the hill base and told the lieutenant that I found a way to get up the hill.

Tired, sweaty and hungry, the lieutenant reluctantly said, "If you think you can make it, go ahead."

There was a little trickle or stream running down the hill, which made a small crevice in the ground between the bushes that lined the hill.

This would be like crawling through a hedgerow, not across it, but completely through the length of the entire row uphill. I got about 12 feet into the hill before I could go no further. My backpack was too big and clumsy and kept catching on the briars and shrub branches and my rifle was always in the way.

I worked my rifle around my body and pushed it down the crevice that I'd just pulled through and told the Marine behind me to have someone bring it up later. I also did this with my backpack that I managed to squirm out of.

Now I had made just enough headspace to crawl up and around the pesky shrub. I kept going, one shrub at a time, breaking branches and twigs just to go a few inches. A couple of times, I thought I would have to turn back, but I kept hearing the troops bitching and moaning behind me. I just got more inspired to press on. I finally found a small clearing about a third of the way up the hill and was able to make good time for a few feet.

But then I was right back at it again, forcing my way up the hill. The flow of water had disappeared, so now there was no trickling stream to belly through. At this point, I didn't care, although I wondered if there was a gook or two waiting at the top of the hill to shoot me.

With gun and ammo somewhere behind me, I really was a sitting duck. After what seemed to be an eternity, I finally saw a clearing at the top of the summit where I could stand for the first time since launching this feat.

The rest of the platoon moved to the open area one by one. They weren't very happy at first. But after the summit was checked out for any other trails, they came to realize that this was probably the safest place they could be right now since it would be pretty hard for the VC to make it up the hill unless they had a secret entry someplace on the hill that was not obvious.

It started to rain, so we cleared the summit the best we could. I helped my comrades make a sleeping area with our flak jackets as pillows and ponchos to keep off the rain. There was nothing to eat so all we could do was tell quiet stories and try to fall asleep. I didn't know who stood watch that night, if anyone. I wasn't chosen due to my resourcefulness to take the hill.

An artillery forward observer was with us so he was able to call the command group and give them a good read of the terrain and what was happening around us.

The morning came, the sun was up and the rain had subsided. The lieutenant got the men going and down the hill we went. With all the crawling up the hill, the path was more accessible. We were on the march again. We were heading north to meet up with the other platoons of India Company.

Along the way, I went to cross a small but deep and rapidly flowing stream and got my boot caught in the crotch of a log I was attempting to negotiate and went head over heels into the rapids.

USMC KRS 1966?

I got a surprise full-body submerged wet down. My wallet, rifle, ammo, backpack and pride all got soaked.

But the dunking to me felt invigorating and I started to laugh at myself. I forgot how tired and hungry I was and just accepted the moment.

About the same time, the lieutenant was coming up from the rear, carrying a marine with heat exhaustion on his back.

I saw an opportunity. I offered to carry the Marine to the medic station if the lieutenant would carry my backpack. The lieutenant didn't know quite what to think, but, because he had to command the platoon, he agreed. I felt good. The lieutenant thought he was getting weight off his shoulders, but wait till he gets a load of this backpack.

I thought!

I put the Marine on my back and off we went. I was happy to help a fellow Marine, but I was also chuckling inside to get rid of that damned backpack. I carried the failing Marine over a mile to a makeshift medic station. I dropped him off and wished him well and was off to catch up with the troops. I got back to the platoon that was mustering in a formation to assault a rice paddy in front of a hill that the VC was entrenched in.

When I got in position, I noticed that my partner on the hill was wearing my flak jacket. It was easy to spot because it had a hand painted Playboy bunny drawn on it. It also had the smoke canister I wore in memorial to the fallen Marine at An Hoa. There wasn't much I could do about it now. I figured I could get it later.

Captain Getlin, of India Company 2nd platoon, told us that we will line up in rows and push out into the paddy with adequate separation so as not to shoot each other.

He also told us if anyone around us got hit, we should not stop, but keep going and the next wave with a medic would provide aid. My lieutenant told our platoon that all we had to do was sweep the rice paddy and take the hill and all the C-rations we could eat would be there for us.

Before the first wave was released, the Navy ship's cannons saturated the area with their artillery, and then the air support came in and did some pinpoint sticks of their own. The mortar team attached to the unit as we moved out, was ready to support any small firearm action.

India platoon was the third wave to go out. The mortars were firing and the enemies were shooting and I was told that there was a gook to my left. I aimed that way and squeezed the trigger, but to no avail. My rifle was jammed. It must have gotten messed up when I was submerged in the stream.

Luckily, the mortar platoon had a seasoned forward observer. He had been out of the Marine Corps for a few years, but re-upped when he learned the Corp needed experienced mortar men in Viet Nam.

The mortar team must have pinpointed the exact co-ordinates of where the gook's nest was because they called in mortar rounds and the enemy fire subsided in that location.

Me and the rest of the platoon pressed on. My mind was numb. I was exhausted, hungry and melancholic. I had lost all fear and purpose and just went on like a zombie. My partner on the hill, who was flanked to my right, screamed out, "I'm hit, I'm hit."

I, without any feeling, said, without looking back, "Just stay there, the next column with a medic will pick you up."

I could hear crying and suffering but moved on with rifles and mortars going off all around us.

One time, the rice paddy was so thick, I just lay down and started rolling to get through the thicket.

On and on we went until we got to the base of the hill.

We got word that it was secured and we proceeded up the grade and, to my surprise, the C-rations were waiting for us. We feasted on the rations like never before. The lieutenant came over and said, "What the hell did you have in that backpack, lead?"

I smiled, "It was just a little handicap I like to take with me on these little outings to make it more of a challenge."

The lieutenant nodded and said, "Sure, next time put a grenade launcher in it," and walked away.

That night, I finally got some quality rest, but, before sleep came,

I witnessed an air show that would make the NYC 4th of July fireworks look like a backyard family picnic. The tracers, explosions, flares and anti-aircraft fire were something hard to describe.

The next day, our platoon was sent to the location where the air show took play the night before. We mustered in a burnt out building. We learned that a big battle had just been completed over the knoll. Some of the troops decided to go up and witness the carnage. I passed on the invitation. I had already heard that it was "blood and guts" all over and it was nothing I wanted to see.

I sat on a stump for a long, long time. I prayed and meditated that all those who had perished and had fallen on these grounds would find their Souls intact and be better for it. It seemed a great price to pay for a higher cause only known by their passing Souls or other Souls of concern. I knew in my heart that the universe was too vast and too unknown for those fallen not to be able to grasp the big picture of what is the common goal of all mankind on Earth.

I wondered if life's experience was set up like a divers meet at the Olympics. Each dive was assigned points for the degree of difficulty and each dive was awarded points for the grace of execution of the dive of choice. Some dives are much more difficult and are assigned more points than ones with a lesser degree of difficulty.

It is the accumulative amount of points by both difficulty and execution that determine if the diver would move up, down or remain the same with respect to his competitors. It also allows the scores to be compared to others throughout the competition during the year or, for that matter, for all who went before them in years past.

I envisioned that a Soul's Olympic Score Card would be set up like this: Souls would select the degree of difficulty that they want to enter human organic life with. They might enter as a rich nobleman with a low score of degree of difficulty, or they might choose to enter life as an orphan child that commands a high degree of difficulty. Execution, however, is different than the degree of difficulty, in that the quality of life and how it is lived is scored the same for all entry levels of life.

The two, degrees of difficulty and execution, are multiplied together, which results in the total contribution score for the life experience.

If the scale for Soul achievement was arbitrarily set at 0-100 and a certain Soul had a universal score of 67, he would need to get at least a 67 to maintain his Soul status. Anything lower would decrease his overall score, any score higher than 67 would raise the universal score. As far as a Soul reaching a perfect 100, it would be extremely difficult because the scale would follow the Law of Diminishing Returns, the closer the score gets to 100, the less scale value it would receive.

I felt that most VC probably had a degree of difficulty that was quite high. If their lives were lived based on their beliefs, then their Souls were enriched from their experience on Earth. Their multiple of degree of difficulty and their quality of life would result in a universal upward score change. They would not be judged by what side of the war they were on, or what side was considered to be more "just" on Earth, their only responsibility is to themselves. The key is for individuals to be able to listen to their inner self and follow their convictions without infringing on any other individual Soul's rights.

The operation was over, and we force-marched back to our An Hoa base camp. Before I marched back, I took the contents of my backpack and dumped it down into the same stream where I took a dunking in. I managed to get through the entire operation without firing one shot or using any of my ammo and still be in one piece.

When I got back, I was reminded that Top wanted to see me. I hastily dressed and rushed over to the headquarters' bunker and announced myself. Top came out of his office, took one look at me and just shook his head. He had never seen a dirtier Marine since he was in Viet Nam. He had gotten used to his desk job and seeing a field Marine fresh out of combat was not part of his daily routine. He instructed one of the PFC's to take me to the supply tent and get me a couple of sets of jungle fatigues. This time, they were not going to be from the "dead man's tent."

I returned that afternoon and met with Top. We had a pleasant verbal exchange. It had been a long time since I had actually gotten to converse without all the Marine Corp hype. We discussed everything from hometown events to governmental philosophy. Top told me that I had the job as a battalion office rat and my duties were to answer the phone on a rotating shift basis during the night. My other major chore was to draw out ambush sites for the troops by using Grid Square co-ordinates.

I was told that the highest-ranking officer and head of the battalion was Major Meadus.

The Major was a no nonsense officer who received a battlefield commission in WWII. The Major was a Marine's Marine. He stood tall, was muscular and his weathered face was complimentary to his staunch demeanor. When he walked into the bunker, all troops and officers came to attention!

I was introduced to the Major by Top and I had all I could do to appear calm and relaxed. The Major welcomed me and went on his way. I was not the most polished trooper in the Corp., but I knew if I wanted to stay with the command group, I'd have to fake it the best I could.

I had phone duty that first night.

The protocol for answering the phone was "fertilize-3" so the caller would know if they got through correctly. I was also given the co- ordinates for the next day's ambush sites. I mapped out the ambush sites through the night with help from two other office rats, PFC Goodie and PFC Pratt.

Goodie and Pratt were considered golden boys for some reason that was a mystery. This OJT training went on for a week. It was dirt simple, but the "crotch" (USMC) likes to do everything in great detail. During this first week, I also needed to get a Marine Corp. driver's license. Top sent me down to the motor pool to pick it up. When I got there, I was sent to see Lt.

Dobbs. Dobbs was a fresh, just out of Quantico officers' training, snot-nosed Second Lewie.

Dobbs told me that I would have to pass a driver's test. He had one of his men take me out to the Jeep and had me drive around with a small trailer attached. All was going well until I had to back up. I couldn't keep the trailer from jackknifing, no matter how hard I tried, with or without the mirrors. I went back to the motor pool tent and Dobbs sent me packing to my bunker with a failing grade. I showed Top the failing report. Top went ballistic, not at me, but at the wet behind the ears Sec. Lewie Dobbs. Top chewed Dobbs a new A-hole. When the air cleared, I had my military license and Dobbs was left with his "gun" in his hand. The next week, I was on my own.

The next day I stood watch without supervision and was given several ambush sites to plot. I did this for three days and then the "shit hit the fan".

Most all of my ambush sites had mistakes in them. I was shocked. This work was so easy I couldn't imagine what happened. I wondered if this was the same problem as when I was in the fourth grade when I thought I aced my math test, only to find out, with the test graded, that I missed several questions that were easy. I wondered when this problem was going to stop jumping up and biting me in the buttocks. Einstein certainly didn't have anything over me when it comes to articulating minute but important details!

Top told me to lay low for a while because Major Meadus was upset. I felt like telling Top, "Who cares? The troops don't follow it anyway." But I didn't. I knew I was responsible but I didn't know why I always made these bonehead mistakes. Top already knew that I typed about as well as a Marine could fart the Star Spangled Banner at revelry, so that was not going to be an option.

I reverted back to my old tactics when all else failed—fake it. I started making people laugh. It didn't matter if they were laughing with me or at me. I just needed to change the air.

51

Whenever I could, I would change an event into a weird situation.

Top must have seen something in me, because in a day or two, I was made the Major's official driver. I couldn't type. I couldn't spell. I wasn't neat or squared away. I couldn't pass his driver's test but Top chose me anyway. I felt like my life was just saved again!

Top let me answer the phone at night, when no one was around.

During the day, I was a gopher, I would have to "go fer" this or "go fer" that for the Major, or anyone the Major decided needed help.

I felt like I saw more action in the first few weeks than most support troops saw in a complete tour.

Our battalion was sent to Okinawa for a 60-day regrouping. We didn't know why, but I knew it was big. We took a ship over and I got very seasick.

The 60-day training was tough, but I liked the great shape I was getting in.

When I did get R & R, I would take a small kamikaze Toyota to town and do shots. Those little Toyota's, or any Toyota, were not common in the US. One time, I went into town with Goodie and Pratt. Pratt was married and often bragged how he was so righteous and would never, ever cheat on his wife. It was amazing what happened next. We hit our first girly bar and Pratt was gone. Some little Meson got a hold of him and off they went. Goodie was a real charmer, so all the Meson girls were seeking him out. I liked the sauce too much and just drank the night away. The 60 days passed very quickly and the battalion was heading back to Viet Nam.

Only this time, we were heading to or near the DMZ.

This is where the fighting was most intense and the enemy was not only the VC but also the NVA (North Viet Army).

When we arrived in Viet Nam, my previous unit India Company, was sent directly to the DMZ.

If Top stuck to his original plans, I would have stayed with India Company as Captain Getlin's personal radio operator, but, as fate would have it, I was put with the battalion commander group.

Captain Getlin, Lt. Bobo and two squads from India Company, 2nd Platoon were sent to Hill 70, which was west of Con Thien in the DMZ.

DMZ

Fifteen good men were lost in battle and 60 more were wounded in an ambush lay out by the NVA. The Hill became known as "Getlin's Corners" where Captain Getlin and Lt. Bobo both perished and were posthumously decorated for valor. Captain Getlin received the Killed in Action-Navy Cross and Captain Bobo was awarded the Medal of Honor.

I, along with the battalion headquarters group, landed south of the DMZ and soon were ordered to Camp Carol. Camp Carol was a camp high on a very large hill and served as base camp for the 9th Marine Regiment Headquarters. It had huge 40 foot Tom Toms (long guns) that could reach far into the DMZ.

Our unit was headed there to provide camp and perimeter security.

As our unit approached the hill, we could see two observation towers made out of huge telephone-like timbers that had to be at least six stories high. There was a small observation deck at the top. I hoped I didn't have to have watch duty up there because I was afraid of heights.

I thought they looked more like aiming stakes for the NVA to line up their mortars to shoot at Camp Carol.

There was an excessive amount of traffic in those next few days because, as our 3rd Battalion moved in, the previous battalion was moving out. The command unit office moved into a scantly sandbag protected tent. I and the other office rats had an unprotected tent just to the left of the command tent. It looked like we would be able to operate there without much trouble.

When we were in Okinawa, Top added another office rat to the crew.

His name was Private Myers and, like me, he was older than most of the privates. We got along great. I didn't have much use for the "Golden Boy" pair, Goodie and Pratt. The command office was still being set up so there was no desk duty. The office rats were left to square away gear in their sleeping tent. Nightfall came and the rats set up a makeshift perimeter watch while the others went to sleep.

About 23:00 that night, all hell broke out.

The NVA were hitting us with everything they had. The NVA knew that the camp was in limbo so they seized the moment.

Myers and I thought we might be overrun so we fixed bayonets and hurried out to the perimeter. The "two golden boys" were huddled in a corner of the tent, scared out of their wits. When the mortar attack died down, they made a dash to the command tent and hid under some sandbags that were next to the sandbag barrier around the tent. They looked like two scared rats that were trapped in a cage. Myers and I finally got the ok that the perimeter was clear and the gooks at the barbed wire were either killed or chased off. We went back to the command tent to see if they needed help there. The two golden boys were still hiding under the sandbags, but the old salts were sitting around telling sea stories from WWII and Korea.

The old salts would sit on unopened crates or sandbags and talk about the old days. When an explosion occurred, they would all hit the deck. Sometimes, sharp metal would rip through the canvas of the tent. The mood was really subdued. It was like everyone was high. The explosions went on into the night, only showing signs of letting up when the sun started illuminating the landscape

One by one, the men started leaving the tents.

I asked Private Myers "where are you going, to the show"? He left the tattered bunker while adjusting the crotch of his jungle fatigues.

"No, I'm just picking my seat for a gourmet dinner at the Ritz!"

With that, we laughed and all day long we lavished over how the two golden twins performed.

The next day, the regiment put on an air/artillery show that would even give Bob Hope a run for his money in terms of glitter and show stopping. They called it a turkey shoot and was it ever! The big Tom Toms shot into the DMZ and the napalm started dropping all around the perimeter, setting off fires and wreaking havoc as they burned the canopy off the jungle for better visibility. They also used Agent Orange to impede the jungle growth.

The camp was never the same. One lieutenant was killed in his rack. He received a direct hit in the initial attack. Many troops were wounded or killed. It was a night few could forget.

The following day our Company had a bucket brigade. The troops lined up at a freshly dropped pile of sand. I grabbed a shovel and started filling sandbags. Pratt was holding the bags. I knew the harder I worked, the harder Pratt would have to work. I worked feverishly. Myers did the same. Goodie was just behind him, so he was getting a grueling workout too. Myers and I were relentless. We were like the Energizer bunny. We just kept going and going.

The golden twins were the opposite. They couldn't keep up and they were always bitching and moaning. The more they bitched, the more intense Myers and my resolve grew.

That night, I estimated that I had filled at least 300 sandbags. I went to the duty log to sign out and noticed it was 21 Apr. 1967.

Twenty-five years before, I was a newborn baby. It was my birthday and I didn't even know it. I took pause to praise myself for hitting the quarter century mark, still alive.

I also thought about the terrible loss of life that I had witnessed and where this whole thing was heading.

The philosophy that I went by was that if the country needed my help, I was there. I heard that back in the states the kids were burning their draft cards. I wondered if they knew something I didn't.

I felt very naïve.

For the first time ever, I questioned my patriotism.

What if the protesters were right and my blind patriotism was flawed. I would be a fake if I thought the cause was flawed. I knew I wouldn't quit anything I stated, but I wondered what I would say if anybody asked me my opinion when I got back to the states.

I spent the remainder of my time with the command group but now I bunked in with the scout troops along with the company "Stars and Stripes" field reporter.

This was a more "settled-in" time for me. I went about my business working for a new major who had relieved my beloved Major Meadus, who rotated back to stateside. I basically did "outpost" watches with the scouts at night and swapped war stories with the company reporter and the scouts during the day. The scouts had a new secret weapon that was not to be discussed. It was a Star Scope that enabled starlight to illuminate the rifle scope lens so we could virtually see in the dark. I used this on my watches.

Being attached with the scouts had its advantages. As a company "grunt" in a field unit or with the Command Group, I was allocated two beers a day. By staying housed with the scouts, there were perks. Since I had practically unlimited use of the Jeep, I would drive the scouts to a nearby village to trade C-rations for beer. I found out this was a common practice by people "in the know" so I felt ok about it. It was also helpful knowing that the C-rations were used by the friendly villagers and the South Viet Nam soldiers.

I was getting near my rotation date (becoming a short-timer) so I started counting the days. But about a week before I was to rotate out, the new major wanted to see the DMZ. We packed up and along with a few scouts, coppered into the de-military zone. During the next three nights my combat experience returned. I realized I had gotten a little soft back at the command outpost.

At the command center I didn't feel surrounded like I was now. The last night there, we were shelled by NVA mortars. Luckily, no one was hit, but it sure was a wake-up call for the Major as well as me.

The Major had us out of there the next morning and I was headed back to the Da Nang airport to go stateside.

The last night of my tour, I was in Da Nang. We were hit with incoming rockets. I didn't want to die on my last day but I had heard and knew of several stories about such incidents.

Upon sunrise the following day, I was still in one piece. To my amazement, I found out that I would be flying out on a commercial Delta carrier. What was even more amazing was that there were American stewardesses on board. After I got on board and was strapped in, I was given a frozen hand towel to cool off with from a knock-out blond who could have passed for one of the Mac Hose girls.

This was one of the most memorable moments of my 25 years and I cherished it from then on.

I've spent the last 20 or so pages giving you a little background about me and my stint in the military which is to show the contrast and similarities of Einstein and me.

Some of our traits are governed by environmental influences, whereas, others are by self-survival due to any learning disabilities we may have, but most likely they were cultivated by both (they were entangled).

Einstein's disdain for nationality probably came from his experiences relating to Hitler.

My strong nationalism (I get goose bumps when I salute the flag and hear the national anthem) is also probably an environmental learned trait relating to my growing up under a stable government established under God.

Einstein's political views about the use of the atomic bomb is as strong as my believing in standing up when called to serve my country.

My military back ground appears to be much different than Einstein's, but maybe not. Einstein was never asked to enlist to fight for a cause

However, he did have to make a morality decision about the atomic bomb. He chose to support the position to stand up against the enemy by use of force and in the long run save lives by shorting the war with the use of the bomb.

As I look back at it now (2014 AD) I believe that I was a patron for volunteering to fight in Viet Nam. I don't begrudge the non-believes that went to Canada etc. to escape the conflict, as long as they felt in their hearts that was right for them.

I don't however think people like Bill Ayers or his wife Bernardine Dohrn, former leader of the radical, anti-war movement Weather Underground, should have done what they did--- right or wrong Viet Nam was a war-- killing cops and blowing up buildings are crimes--there is a difference.

Sharon and I, some 50+ years after my High School Graduation. She has five beautiful children, eight grandchildren and two great grands that are all adorable. I have one son who I love very much, so all in all I have no regrets with respect to how life has progressed so far!

OOPS, I GOOFED.... Hear it is!⏃

Photo taken by Richard Corrsett, a college buddy and a good friend.

Part II

Creative Thoughts

"Imagination is more important than knowledge."

"Any intelligent fool can make things bigger, more complex, and more violent. It takes a touch of genius -- and a lot of courage -- to move in the opposite direction."

Before we look at how Einstein used his creative juices to change the way people looked at the universe, I want to first review some fundamental points about; size, dimensions, space, time, and the speed of light. Also dark energy (De) and dark matter (Dm) are included. Some of these fundamentals were developed after Einstein's era. Along with these fundamental concepts I have also interjected some of my own creative thoughts about what I hypothesize.

These fundamentals all point to my thought experiment which results in ~ 10^{183} synchronized Planck size, space, time, zero- point-energy particles with eleven dimensions. I refer to these particles as Space-Time-Energy Particles (STEPs). This is a new way of looking at our universe and is the result of creative thinking backed up by referenced knowledge.

The very first specialized embryonic STEP is the space-time-energy-nugget (STEN) that rendered all additional nuggets/particles.

Size

When in High School you were probably introduced to Avogadro's number being ~ 6.22×10^{23} particles in one mole of a substance (incidentally this is a number that Einstein worked on). This is a large number for very small particles and is very hard to imagine Actually if you think of a particle being a string with a length of one Plank's unit (~1.62×10^{-35} meters), then we are

61

talking about an absolute length at the limit of our measurement scale at the lower end. Any length smaller is considered to be outside the scope of our mathematical comprehension! In terms of a quantum upper limit size of a mega universe, the number is 10^{500} different universes. This is a mega universe number discussed by string physicist such as Leonard Susskind.

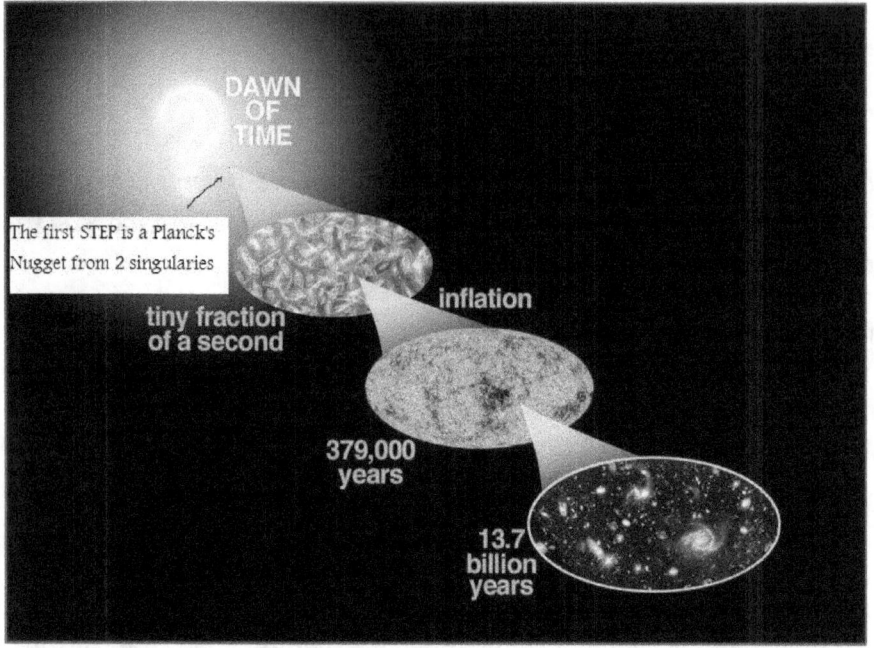

This is a quantum derived number and is considered our upper limit for a universal finite limit.

 Fig# 01 NASA, Universe from a Planck's nugget (STEN) $\sim 10^{-35}$ m to the observable universe (OU) $\sim 10^{27}$m.

The current diameter size of our observable universe (OU) is 10^{27} meters. According to the big bang theory and the inflation theory it was only the size of a cubic Planck size nugget, space-time energy- nugget (STEN) when it burst into existent.

http://htwins.net/scale2 is a perfect site to grasp size!

To visualize these huge numbers, think of an 8 oz. glass of water that has ~10^{24} molecules of water in the glass. The total number of 8 oz. glasses of water on earth is ~10^{21}, including all the oceans, lakes, rivers, atmosphere etc.

Comparing these numbers, there are about 1,000 times (10^3) more molecules in a glass of water than there are glasses on earth.

Now, let's suppose that Einstein drank that glass of water then redistributed the 8 oz. back to the bulk of all the water on earth. Over time all of these molecules distribute evenly throughout the total water bulk. After this dispersion another 8 oz. is drawn to quench your thirst.

What is the probability that you will consume some of the same water molecules that Einstein did?

It is likely that your glass will share some of the molecules that were in Einstein's glass.

Why?

Because there are 10^3 times more molecules in one (10^0) glass than there are equivalent glasses of water on earth.

On average, your glass will contain about 10^3 molecules that were also in Einstein's glass. This is true for any two (2.0×10^0) glasses.

One thing that I'd like you to take away from this is the relationship between your glass of water and the number ~10^{21} of glasses of all the water on earth.

So when you read about very small numbers or very huge numbers, you can refer to your glass of water to all of the water on earth.

"Not everything that counts can be counted, and not everything that can be counted counts." (Sign hanging in Einstein's office at Princeton)

One very important concept I'd like you to bear in mind relates to size. Remember that infinity never has a beginning and never has an ending.

Whereas finite has a beginning and an end. Throughout this writing I consider the universe to be finite and only immorality of the human Soul and Eternity are considered in this writing to be infinite.

"Two things are infinite: the universe and human stupidity; and I'm not sure about the universe." (Seems I'm more sure about the universe than Einstein)

One of my pet peeves is when theoretical physicists try to push their math to a limit of zero instead (for example) of a Planck's limit of~10^{-35} meters on the small end and/or infinity at the upper end, where I propose the number to be ~10^{500} units (the mega verse limit) which is what Quantum theory predicts.

"Whoever undertakes to set himself up as a judge of Truth and Knowledge is shipwrecked by the laughter of the gods."

I also want to point out that, in this book I refer to energy and matter often. Keep in mind that normal energy (Ne) and normal matter (Nm) can be exchanged by Einstein's equation, $E=mC^2$, $\{Ne=NmC^2\}$.

When I discuss a STEP as a STEN (Space-time-energy-nugget) I am actually referring to a special case of a STEP that has an abundance of dark energy (De).

Abundant dark energy concentrated in a STEP is defined as a STEN and the abundant dark energy is what dark matter (Dm) is. This is similar to the concept that normal matter is just concentrated normal energy, only here $De=DmC^2$.

I am getting ahead of myself, just bear this in mind when we discuss the STEN.

I introduce size and dimensions here because this plays a very important part in the understanding of how the observable universe works.

At the lower end of the scale from ~ 10^{-35} m to $10^{-16.5}$ m we are not able to make any microscopic observations. Only mathematical computations suggest sizes smaller than ~$10^{-16.5}$ m.

This will become very important when we discuss kinetic energy, standing and travel waves, Heisenberg's' uncertainty principal and other quantum concepts.

The observable universe (OU) as shown in Table#1 is ~10^{27} meters in diameter (or when considering the OU to be cubic then it's the length of its sides).

This is the current OU size, it does not project any dimensions beyond the event horizon of the OU.

Dimensions

Orders of magnitude (length)

From Wikipedia, the free encyclopedia, a modified table of Examples of the orders of magnitude for different lengths.

Section	Range (m)		Unit	Example Items
	≥	<		
Planck length	?	10^{-35}	ℓ_P	Quantum foam
Subatomic	10^{-35}	10^{-15}	am	electron, quark, string
Atomic and cellular	10^{-15}	10^{-12}	fm	atomic nucleus, proton, neutron
	10^{-12}	10^{-9}	pm	wavelength of gamma rays and X-rays, hydrogen atom
	10^{-9}	10^{-6}	nm	DNA helix, virus, wavelength of optical spectrum
Human scale	10^{-6}	10^{-3}	μm	bacterium, fog water droplet, human hair's diameter[1]
	10^{-3}	10^{0}	mm	mosquito, golf ball, domestic cat, violin, football

	10^0	10^3	m	cello, human being, automobile, sperm whale, football field, Eiffel Tower
	10^3	10^6	km	Mount Everest, length of Panama Canal and Trans-Siberian Railway, larger asteroid
Astronomical	10^6	10^9	Mm	the Moon, Earth, one light-second
	10^9	10^{12}	Gm	Sun, one light-minute, Earth's orbit
	10^{12}	10^{15}	Tm	orbits of outer planets, Solar System
	10^{15}	10^{18}	Pm	one light-year; distance to Proxima Centauri
	10^{18}	10^{21}	Em	galactic arm
	10^{21}	10^{24}	Zm	Milky Way, distance to Andromeda Galaxy
	10^{24}	$10^{10^{10^{122}}}$	Ym	Huge-LQG, Hercules-Corona Borealis Great Wall, visible universe, quantum multiverse

Table 1. The size of the visible universe is reported to be between 10^{24} and $10^{10^{10^{122}}}$. The visible universe is also referred to as the observable universe (OU) and is shown to be 10^{27} meters in diameter. I will use this number 10^{27} for the size of the observable universe (OU) in this book as I relate it to events such as the big bang, inflation or expansion etc.

The size of the observed universe (OU) is not static since Dark Energy continues to expand matter- antimatter farther and farther away from the original origin (the plank's embryonic nugget, STEN) of our universe.

Often Quantum theorist use the number 10^{500} instead of $10^{10^{10^{122}}}$ for the total number of   other universes (multiverse) besides ours. This size will also be of importance later.

I would like take the position that " it's not size that matters," it's the understanding that the universe(s) is/are finite and only the singularities of Eternity or of Souls are infinite (meaning don't get blown away by the huge numbers that we'll deal with).

Time

There are ~10^{43} Planck ticks of time in every second we experience.

There are more units of Planck tics in one second than there are seconds since the big bang. There have been 10^{17} seconds since the big bang.

To get a feel for this short time span, place an object in a microwave and wait to the last second to show up on the digital panel. See how long you can read the one second digit before the alarm goes off.

During this time span 10^{43} Planck ticks are possible. If you beat the buzzer by opening the door before it goes off, how many Planck ticks did you experience?

Planck's time is the time it takes for light to travel Planck's length. Actions across lengths less than this boundary have no meaning.

The smallest length (Planck's length) divided by the fastest speed (the speed of light), is the time it takes for the fastest thing to travel the shortest distance, 10^{-35} meters/10^8 meters/Sec. = 10^{-43} Seconds.

Thus, times shorter than Planck's time do not make sense. Conversely if you divide the length of a Planck's unit by the number of ticks in a second you get the speed of light. 10^{-35} meters/10^{-43} seconds = 10^8 meters /seconds.

Dimensions & Time (The first four dimensions of the Universe)

We normally think of our world to have three dimensions: H=height, D=depth and L=length, plus the 4th dimension of space-time. I would like to offer another way to look at the first four dimensions of our universe plus the additional dimensions of the M theory that was developed by Dr. Edward Witten.

Let's first look at our universe when it was only Planck's size. A Planck size nugget, a STEN.

A STEN is a Space-Time-Energy-Nugget which is a specialized case of a Space-Time-Energy-Particle, a STEP. Again, I'm getting ahead of myself, so just hang onto this concept.

A Planck size nugget is a cubic nugget with a cubic size of 10^{-35} meters cubed (10^{-105} m^3).

Two of the sides of the nugget are parallel squares and are considered to be two-dimensional space-membranes (quantum D-brane), each with a height and depth (HxD) of one Planck's unit square. These two squares represent space-membranes. Their two dimensions, H&D are the first and second dimensions of the universe.

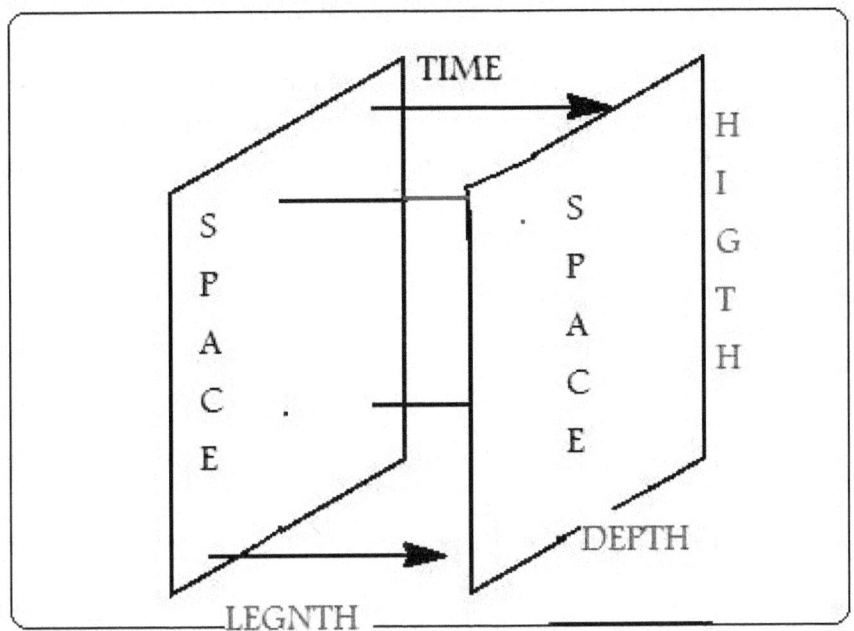

Fig# 2A. Space membranes (a quantum D- brane) has square dimensions of one Planck's unit squared and is the first two dimensions of our universe. These two dimensions are height and depth.

Space Time

Space-Time is at least four dimensional.

If we introduce the speed of light or speed of any electromagnetic wave in a vacuum perpendicular to the two space membranes, and the distance traveled between both membranes is one plank's length, then the distance it travels (1 Planck's length) is the third dimension and the time it takes to go that distance is the 4th dimension. To sum it up, we have in a vacuum:

1. Space membranes = Planck length (L) squared or ~10-35 m^2, which represent the Height and Depth of space membranes.

2. Speed of energy in a Planck's cube = Planck's L~10^{-35} m / Planck's time (T) ~ 10^{-43} sec =~ 10^8 m/sec.

69

Note: ~ is an approximation symbol used by mathematicians.

3. This suggests that there is energy in space/time, not the kind of normal energy that relates to the conservation of normal matter, but to dark energy, which to date is not understood completely. Dark energy is still

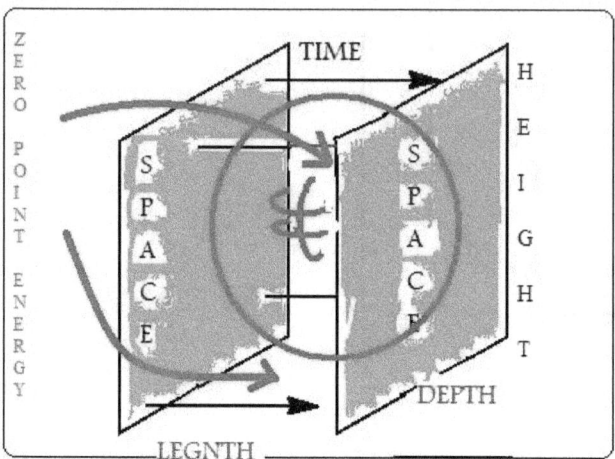

considered a mystery as to its origin and operandi.

Fig# 02B. Space membranes perpendicular to time /length with zero-point energy (dark) sandwiched into a Space-time-energy-particle (STEP).

Zero Point Energy (ZPE)

The concept of zero-energy point was perceived in Germany by Albert Einstein and Otto Stern in 1913, using a formula developed by Max Planck in 1900.

Vacuum energy is the zero-point energy in the cosmos. Vacuum energy is one possible explanation for the cosmological constant.

I hypothesize that vacuum energy is dark energy (De), or concentrated dark energy which I'll explain later to be dark matter (Dm).

Dark Energy is found between the two parallel space membranes as equal, but opposite opposing forces that net zero normal energy due to the cancellation of each other, much like when two waves cancel each other out when perfectly aligned.

The two Zero-point apposing dark energies that fit inside the first original STEP is a special case STEP because it's dark energy was so concentrated that it was actually dark matter, so in this case I consider dark energy and dark matter as interchangeable commodities. I refer to this special case STEP as a Space-Time-Energy-Nugget (STEN). The N (Nugget) implies that the STEP is a specialized STEP that has dark energy concentrated as dark matter. The two opposing energies manifest themselves as two tesseract manifolds. One is cubic (Fig#03) and the other is cubic on a 3-sphere (Fig#04).

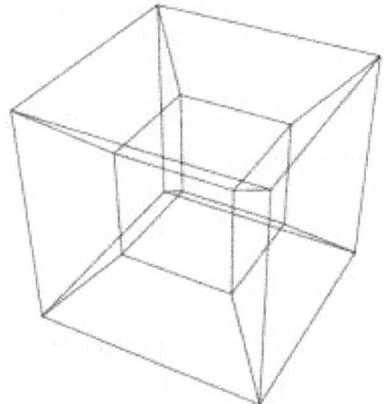

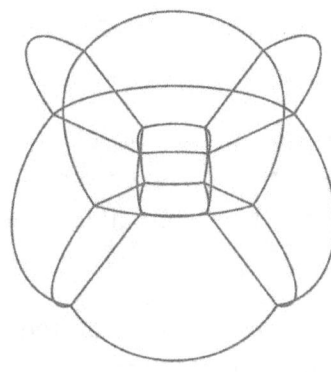

Fig#3. Cubic tesseract   Fig#4. Cubic tesseract on a 3-sphere

Initially the ZPE was 100% DM. The dark matter of the 3-sphere has a larger Volume than the alternating DM cubic tesseract, so it did not fit completely into the embryonic Planck's cubic nugget (Fig#05A).

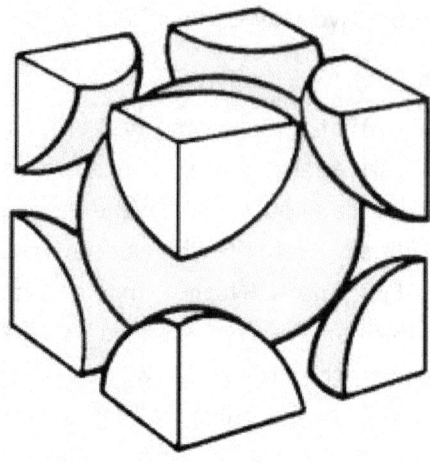

Fig#5A. The volume of a Sphere inscribed in a cube is 52.38% and the remaining volume is 47.62%.

The Eternal force (called the Super force by others) that spurred the Big Bang used its power for the first time in the new universe by reducing the volume of the 3-sphere from 52.38% to 47.62%, resulting in a phase shift, making the two alternating tesseracts super symmetrical with respect to volume.

To accomplish this some of the Dark matter (Dm) congealed into normal matter (Nm), by an Einstein like equation. **Dm= [De − Nm]/C^2, where c is the speed of light**

The Eternal force still had plenty of power stored in the dark matter of the STEN. The phase change was the volume correction. The volume of the tesseract on a three sphere was reduced by 4.76 % by condensing this amount of dark energy/matter into normal matter fermions and Higgs/force bosons along with ghost/shadow Sparticles. This created the wormholes of the tesseract on a three sphere which now can be referred to as a six dimensional Calabi Yau tesseract which increased the dimensions to 10. The 11th dimension came earlier by way of an onion skin condensation around the 3-sphere resulting in a graviton boson. See fig# 5B for as illustration of A STEP/STEN.

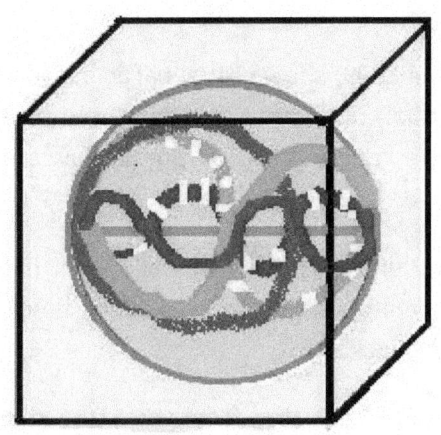

STAUFFER'S
SPACE-TIME-ENERGY-PARTICLE
STEP

Fig#05B.The tesseract 3-sphere has 6 worm holes and an onion skin in between the tesseracts (Calabi-Yau). The total dimensions of a STEP are 11 counting the 2D space membrane and the meters per second of the alternating dark energy/matter.

This first embryonic STEN reproduced its self, 10^{80} times.

These 10^{80} STENs in-turn produce 10^{60} STEPs as shrouds around each STEN. When the ZPE was first formed by the big bang it was completely consolidated in the first specialized STEP, the embryonic STEN a (Space-Time-Energy-Nugget).

The STEN is identical to the STEP except that it contains concentrated abundant dark energy which as stated before is dark matter. Actually as stated above, the STEN is a specialized STEP. This may sound redundant but it is a very important component in my theory of the "Grand Original Design." And the Holy Grail Equation of physics.

The STEP is the opposite of the specialized STEP/Stern in that, it is mainly composed of dark energy.

Feynman diagrams (See pages 24 & 99) show virtual particle involvement in entanglement, etc. These virtual particles are considered to be Dark Matter. However, virtual particle violates classical physics law and are not considered here

Virtual particles are replaced by "time channels" that allow normal- matter-string- particles to travel as traveling waves. The residence time in a time channel is 10^{-43}seconds, or one Planck's tic.

In summary; the STENs can be differentiated from the STEPs due to their dark matter content.

Additionally, STEPs, produced from STENs, expand the observable universe to its present size of 10^{27} meter in diameter or cubic side length.

These newly formed STEPs are not detected by gravity due to their lack of any appreciable (if any) dark matter.

When STENs produce STEPs, the universe not only expands, but the dark matter decreases as the dark energy increases. (See Fig#06A &B).

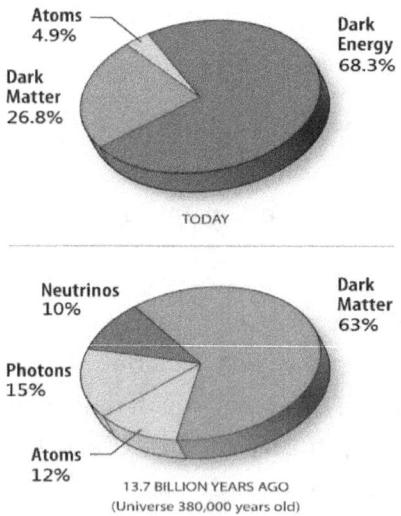

74

Fig#06A.This NADA pie graphs show the universe when it was 380,000 years old and also of today. It suggests that as dark matter decreases dark energy increases.

Warning! This graph will continue to change slightly as better and better data is collected this image is made with 5 year WMAP data. The final 9 year data produces a more accurate result.

WMAP data reveals that its contents include 4.6% atoms, the building blocks of stars and planets. Dark matter comprises 23% of the universe. This matter, different from atoms, does not emit or absorb light. It has only been detected indirectly by its gravity. 72% of the universe, is composed of "dark energy" that acts as a sort of an anti-gravity. This energy, distinct from dark matter, is responsible for the present-day acceleration of the universal expansion. WMAP data is accurate to two digits, so the total of these numbers is not 100%. This reflects the current limits of WMAP's ability to define Dark Matter and Dark Energy.

Credit: NASA / WMAP Science Team.

Content of the Universe - Pie Chart

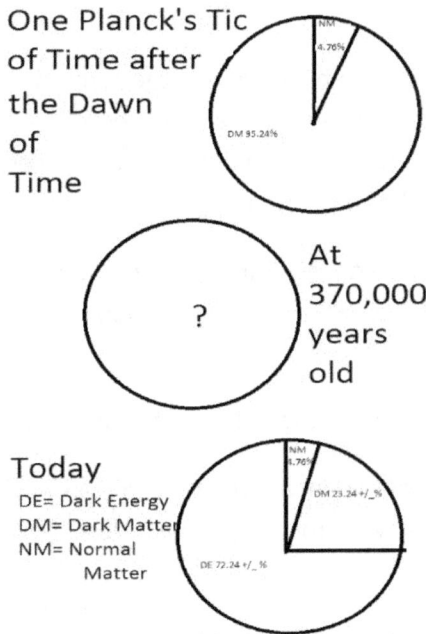

One Planck's Tic of Time after the Dawn of Time

NM 4.76%

DM 95.24%

?

At 370,000 years old

Today
DE= Dark Energy
DM= Dark Matter
NM= Normal Matter

NM 4.76%

DM 23.24 +/_%

DE 72.24 +/_ %

Fig#06B. The way I think the OU transformed as seen here is that after Normal matter was introduced into the (OU) it always maintained its 4.76% of composition. However, the ratio of DE to DM change. As the DM decreased the DE increased. As to the 380,000-year-old pie graph of NASA---I'm not sure how this fits?

Einstein referred to this vacuum Energy (dark energy) as a cosmological constant to combat gravity and keep the universe in a steady state as opposed to a dynamic state.

As it turns out, this constant causes the universe to expand (actually Einstein waffled on this issue) since steady state/dynamic principles were not at play until Edwin Hubble's epic discovery. Hubble first reported, in 1929, of an expanding universe where the speed at which galaxies receded from our point of view was directly proportional to their distance, a relationship we now call Hubble's' Law.

http://oneminuteastronomer.com/5576/expanding-universe#sthash. lvm90daD.dpuf

The reason that the new STEPs are not affected by graviton's the same way their precursors are is similar to the reason that some photons do not extract electrons from their location and some do, (Einstein's Noble Prize Photo electric effect, {see notes at the end of the book}).

Here the newly formed STEPs do not have enough Dark Matter (if any) in their manifolds to be attracted by gravitons, whereas the precursor STENs do.

This will become clearer when I address the big bang followed by inflation.

A metaphor to explain the difference between a precursor STEN with abundant Dark Matter to that of a newly formed STEP with less Dark Matter (if any) is as followers:

1) Push both of your palms toward each other and notice the increase in force on each palm. This is an analogy of a precursor STEN with an extensive amount of dark matter.

2) Place the palms of your hands against each other, without pressing and feel the force (or lack of) against each other. This is an analogy of a newly formed STEP with a small amount (if any) of dark matter.

The Eternal force is contained in the dark matter, and only when it is completely exhausted will the universe stop expanding. The Eternal force is the force that accounts for the tesseract rotations and switching of position and geometry of the two Calabi Yau tesseract opposing energy/matter that neutralize each other as ZPE.

They change positions and geometry every 10^{-43} seconds. All STEPs (10^{183}) are synchronized, similar to the way the cells in our body are.

The Eternal force Ef is equal to dark matter Dm times the cosmological acceleration Λ_a when time T and temperature Θ are implemented!

$$[Ef=Dm\Lambda_a] \, T, \, \Theta$$

As equal volumes of dark matter from STENs (Dm=De/C^2 {see Part VII}) leak out of their respective manifolds, they produce dark energy that expands the universe outward. Each expansion by these measured leaks result in newly formed STEPs. These newly formed STEPs are identical to their precursor except they are void of any extra dark matter that would produce additional STEPs.

5th through 11th Space-Time Dimensions

Now that I have touched on ZPE/ZPM found in STENs and STEPs, I will address the remaining dimensions of the universe.

The first four dimensions of STENs, containing dark matter is a duel dark matter system netting in zero-point dark matter, by offsetting of each other. Much like that of two waves canceling each other out as discussed earlier.

This duel dark matter system that neatly fits into space-time was not that way at the dawn of time (the first Planck tic). Earlier (see Fig#05A & 05B), I hypothesized that dark matter converted to normal matter to allow the sphere's volume too equal that of the cubic volume with the four sides of the tesseract cube to be equal to the diameter of the sphere.

When the sphere is reduced to 47.62 % volume by shaving it ever so slightly and drilling several cork screws, and other worm holes through or inside it, the sphere and the cube became identical in volumes.

This was done by congealing 4.76% of the sphere's volume Dark matter into Normal matter (fermions and bosons,) creating the wormholes etc. to exist as dimensions.

Before I expound on this, I want to consider symmetry and geometry of the two ZPM/ZPE manifolds found in the STENs/STEPs.

Consider the opposing Dark Matters to be a cubic tesseract and a tesseract on a three sphere, where both have equal volume's as just explained (see Fig's #03 and Fig#4). This satisfies the ZPM/ZPE and symmetry requirement of duality.

However, to be completely symmetric both would have to be either cubic or rhombic, but since one is on a three sphere and the other is not, a symmetry break must have occurred at the initial Planck's tic of the big bang, where two singularities entangled.

Symmetry breaking is a reduction in the amount of uniformity a system appears to have and is usually associated with a phase change. In this case the phase change was the big bang (two singularities interacting) and the reduction of symmetry was the addition of a three sphere change to one of the tesseracts. This arrangement of two tesseracts might be considered as a mirrored pair. You might ask "How could a cube and a sphere be mirror images of each other"?

Simple, if the mirror is a spherical ball then the cubic image projected on it will still be a cube but project as a sphere.

Additionally, we will consider the two tesseracts as two Calabi-Yau manifolds equal in every way except for their projection. They exchange positions in the matrix of the one cubic Planck nugget every 10^{-43} sec. They are separated by an onion skin (see Fig#05A and Fig#05B).

This exchange of positions of two dark matter manifolds are driven by the eternal force.

The two alternating tesseracts were developed by mathematicians to represent a 4-dimensional cube and a 4-dimensional cube on a 3-sphere. However, the ZPM in the STEN actually have 6 dimensions and should be represented as two (six-dimensional) alternating Calabi-Yau manifolds. Note: only one set of six dimensions are available at any 10^{-43} sec. because as only the 3-sphere arrangement allow for any openings.

Each of the six dimensions of the Calabi-Yau manifolds relate to six worm holes that allow matter, string- particles (fermions) and force, string-particles (bosons) to move through the STEP/STEN along with Sparticle-ghost shadows.

A Calabi-Yau can have thousands (10^3) and thousands of geometric solutions, and to this point in time no physicist has as yet picked one.

In my scenario I select a Calabi-Yau as a cubic tesseract alternating with another Calabi-Yau tesseract on a three sphere. The spherical manifold has 6 wormholes and an onion skin separator between the two manifolds.

The additional six worm holes plus the onion skin added to the first four dimensions' accounts for all 11 dimensions.

The six worm holes and onion skin carved out of the manifold on a 3-sphere account for the 4.76% reduction in volume to give equal volumes to both Calabi-Yau. One wormhole is axillary, four are corkscrews around the axis, and another is a ringworm.

The onion skin was discussed earlier.

The particle zoo will further explain the mechanism of how this all took place during the first few ticks of time.

Fig#05B show the 5-11 extra dimensions of space-time, 4 corkscrews, 1 axillary worm hole, 1 ring wormhole and 1 onion skin dimensions are depicted as ordinary objects.

The 5th through 10th dimensions of space-time are channels that allow normal particles (for example Fermions and Higgs Bosons) to pass through them with spins of 1/2 and 0 respectively. Also the Sparticle-ghost shadows are found in the 3/2 twisted ring which is one of the lesser understood dimensions that I postulate.

This was a little redundant, but needed repeating for additional clarity.

The 11th dimension is the result of the creation of the onion skin that separates the two tesseract Calabi yau's.

Fig# 07. A six dimensional Calabi-Yau manifold represented by Wikipedia

Part III

Einstein got it Right

"The eternal mystery of the world is its comprehensibility."

"Everything should be made as simple as possible, but not simpler."

Einstein got it right about the speed of light!

There have been some recent concerns about the speed of light not being constant as quoted in a BBC update on 30 March 2012;

"Prof Antonio Ereditato oversaw results that appeared to challenge Einstein's theory that nothing could travel faster than the speed of light.

Reports said some members of his group, called Opera, had wanted him to resign.

Earlier in March, a repeat experiment found that the particles, known as neutrinos, did not exceed light speed.

When the results from the Opera group at the Gran Sasso underground laboratory in Italy were first published last year, they shocked the world, threatening to upend a century of physics as well as relativity theory - which holds the speed of light to be the Universe's absolute speed limit.

The experiment involved measuring the time it took for neutrinos to travel the 730km (450 miles) from Cern laboratory in Geneva, Switzerland to the lab in Italy.

Speaking at the time, Professor Ereditato added "words of caution" because of the "potentially great impact on physics" of the result.

"We tried to find all possible explanations for this," he said.

"We wanted to find a mistake - trivial mistakes, more complicated mistakes, or nasty effects - and we didn't."

"When you don't find anything, then you say 'well, now I'm forced to go out and ask the community to scrutinize this."

"Despite the call for caution, the results caused controversy within the world of physics."

If the findings had been confirmed, they would have disproved Albert Einstein's 1905 Special Theory of Relativity.

Earlier this month, a test run by a different group at the same Italian laboratory recorded neutrinos travelling at precisely light speed.

Sandro Centro, co-spokesman for the Icarus collaboration, said that he was not surprised by the result.

"In fact I was a little skeptical since the beginning," he told BBC News at the time.

"Now we are 100% sure that the speed of light is the speed of neutrinos."

So far, Professor Ereditato has not commented on his decision to step down from his post.

http://www.bbc.co.uk/news/science- environment-17560379)

I have also thought about light or all electromagnetic waves moving through space-time as travel waves by means of photons as Einstein brilliantly described and received the Nobel Prize for.

In a book I wrote "Infinity of the Soul" I used an illustration of two cars traveling from a parking lot to home. Car A was on a straight road while Car B took a curved road home.

I took pause for a moment and thought deeper about light being a particle (photon) as well as being a wave at the same time. I took out a 3x5 card and wrote on the front and back:

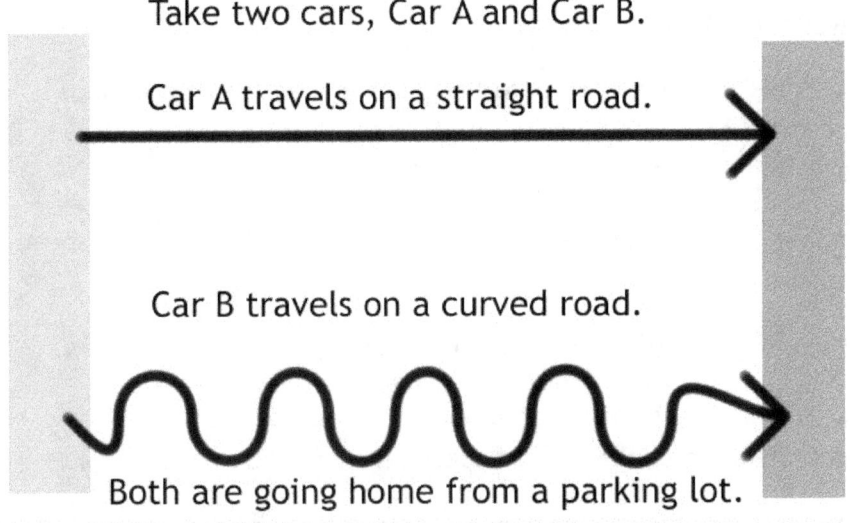

Take two cars, Car A and Car B.

Car A travels on a straight road.

Car B travels on a curved road.

Both are going home from a parking lot.

If Car A travels at the speed
of light to get home,
how fast would Car B
have to travel to
get home at the same time,
assuming they both left at the same time?

Fig#

08.

: From this car-road analogy for photon-wave-travel, I concluded:

"A photon travels faster than the speed of light, but moves through our Universe at the speed of Light."

It is only recently that I've realized why this is possible.

The photon does not travel faster than the speed of light while negotiating through space-time.

When traveling perpendicular to the space membranes, a photon is going at the speed of light, but when it moves parallel to space it is simply in the time wormhole where space is not involved.

Therefore, this analogy does not violate Einstein's principal of the speed of light in a vacuum.

Fig#09 is an illustration of the motion of a photon wave cycle.

When the photon is moving perpendicular to space, it travels at the speed of light (meters/sec.) through the space-time wormhole.

When the photon moves parallel to space, it is in the time wormhole and has no space association. You might say it has superluminal motion but it is hidden inside Planck's time.

In our world nothing can travel faster than the speed of light. The proper statement about the post card should read:

"A Photon travels at the speed of light through space-time, but when moving through time it has no restriction where as a car is always traveling through space time so in my analogy it would be impossible for car B to reach home because it would have to travel faster than the speed of light."

If on the other hand I drew an equivalent path for photon- light, both paths would have photons reaching home at the same time. This is why light waves with different wave frequencies both travel at the same speed, the speed of light.

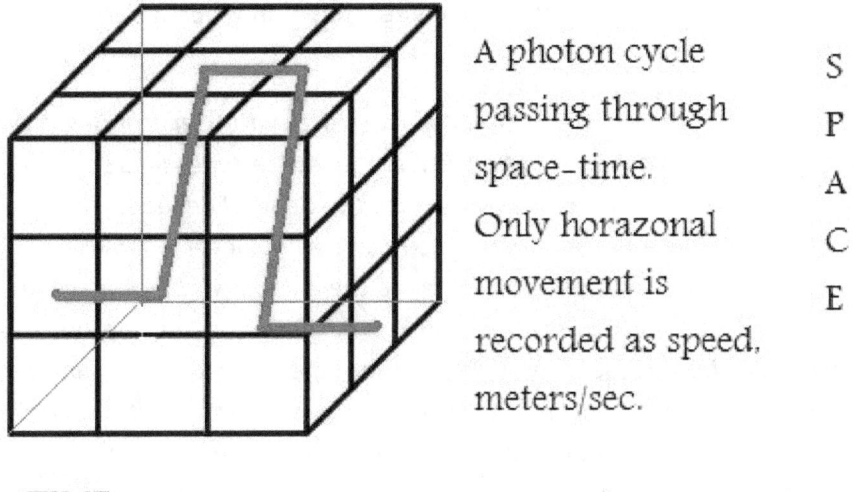

A photon cycle passing through space-time. Only horazonal movement is recorded as speed, meters/sec.

S
P
A
C
E

TIME

Fig#09. A photon traveling at the speed of light must pass through space perpendicular membranes to it when its speed is measured. Time corroders (Time-Channels) are irrelevant when space-time travel is calculated. Also see Fig#25.

This concept may seem like a minor detail, but in actuality it relates to a basic fundamental of our universe in that "Our universe is currently composed of ~10^183 STEP's (and growing) and are synchronized with ZPE tesseract motion to allow normal matter and force particles too transverse throughout its vast volume as one harmonious orchestrated symphony."

Einstein got it right about general relativity

Brian Green in his book The Elegant Universe wrote *"Einstein once again revolutionized our understanding of space and time by showing that they (time and space) warp and distort to communicate the force (of gravity)."*

Another way to look at this is to understand that "Gravity's attempt to pull all of the universe's volume back to the singularities from which it came will invariably warp and distort space-time throughout its journey until even

85

space its self is returned to its original size, shape and location, rendering time a non-factor!"

"Loop quantum gravity on the other hand, starts with General Relativity and attempts to create a quantum version of space-time. It argues that space-time is comprised of loops upon which the General Relativity metric is defined. Outside the loops is nothing - no metric whatsoever. Loop quantum gravity has successfully been used to derive the entropy of black holes, but so far falls far behind string theory in it's all encompassing nature (for example how to represent particles). String theory and loop quantum gravity are both trying to achieve the same goal - quantize gravity and unify General Relativity and quantum mechanics into a seamless theory of everything. However, they are doing it from opposite direction- string theory starts in the realm of quantum mechanics and tries to introduce gravity as just another string (the graviton). Loop Quantum Gravity tries to quantize gravity (and thus space-time) using General Relativity as a starting point, and presumably hopes to try to introduce the features of quantum

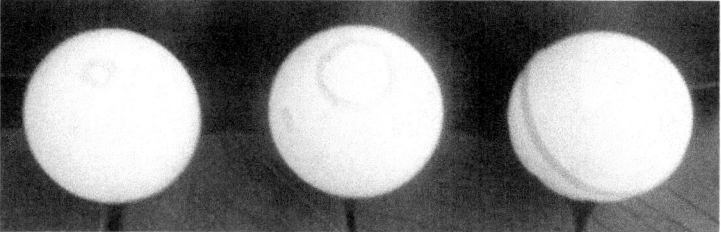

mechanics".

Fig#10. The 2-spin Graviton in the onion skin is only detected at its maximum when it is fully extended and is one Planck's unit long when found at the equator of the sphere.

My view is to consider the loop quantum gravity from Einstein's General relativity to be the closed-in loop.

This is a good starting place to merge both theories together.

Since general relativity has space-time dictating what gravity is, it is logical that when some of the Dark Matter/Energy cooled and congealed that a graviton was formed when the onion skin (11th dimension) was created.

The graviton congealed as a 2-spin closed ring (similar to an ordinary O-ring) string (see Fig# 10).

The closed loop has a diameter of one Planck's length when in the quantum region of space-time that is found between the two tesseract of space-time-energy where the onion ring volume is found. However, this diameter is only true when the ring is at its maximum; see Fig#10 again.

Gravity whether its micro or macro operates by one simple mechanism. The graviton travels at the speed of light and has a single purpose of pulling all matter-antimatter back to its original site, (the original singularities) as discussed earlier. In keeping with Einstein's general relativity, it does so by operating in the space-time onion skin.

The concept of a graviton, a 2 spin boson, being hidden (not being at the equator of the cubic Calabi Yau on a 3-sphere most of the time) supports the parallel theory of Lisa Randall where she claims that the graviton spends more time in a parallel world than in ours and therefore making it much weaker than it seems.

In this scenario the 2-spin boson graviton operating in the onion skin is in the parallel world which only allows the graviton to pull when fully expanded.

The interior of a STEP can be looked at as a parallel world since its interior is hidden from ours.

The STEP can also be considered to be looked at as a complex wormhole but not a black hole since it does not collect matter. However, if a STEP is lacking a white hole then a black hole is created. The white hole is simply the opposite end of a wormhole(s) from where string matter makes entry.

Einstein got it right about Black Holes

When Einstein introduced the concept of black holes there was a lot of misconceptions and confusions by the public as well as the scientific community about them.

Currently a lot of information about black-holes has been worked out. Leonard Susskind and Stephen Hawking have an ongoing debate about black holes as to whether or not information was lost in a black hole.

By the use of a hologram analogy Susskind settled the issue to explain that information can be stored on 2-dimensional space [membranes].

{Note; you might say that he was referring to the STEP(s) that I've been referring to.}

Stephen Hawking showed that black holes emit radiation but the radiation did not allow the incoming info to be regurgitated (it was lost) which violates the quantum theory that information can never be lost.

In the universe there are black holes and super black-holes plus there are the quantum black holes that represent the smallest size for any black hole.

I view these Planck size black holes to be STEP(s) missing the white hole (exit port) in their geometry.

Closed end worm-holes without a white hole exit cause matter to stall and concentrate mass into a confined location. This mass only attracts more mass to the site. For example, a photon's standing wave occupies 10^{60} STEPs. If a photon finds its way to the event horizon it would shed its standing wave STEPS and settle as a single STEP at the bottom of a multiSTEP black hole with its history written on the space membranes of its encasement. It would stay there until it finally reached its destiny as a part of a recombined singularly.

However, if it did evaporate it would do so as an encased proton with history intake.

Since the famed debate of Hawking and Susskind about black hole integrity another storm has erupted about the presents or lack of a firewall in a black hole.

Below is an italicized excerpt from HTTP (physics) that outlines the current situation. I will comment on this with non-Italicized views by use of Bold Lettering.

The motivating paradox

A single emission of Hawking radiation involves two mutually entangled particles. The outgoing particle escapes and is emitted as a quantum of Hawking radiation; the in falling particle is swallowed by the black hole. According to widely accepted research by Leonard Susskind and others, the outgoing particle must be entangled with all the Hawking radiation the black hole has previously emitted. This creates a paradox: a principle called "monogamy of entanglement" requires that, like any quantum system, the outgoing particle cannot be fully entangled with two independent systems at the same time; yet here the outgoing particle appears to be entangled with both the infalling particle and, independently, with past Hawking radiation.

In order to resolve the paradox, physicists may eventually be forced to give up one of three time-tested theories: Einstein's equivalence principle, unitarity, or existing quantum field theory.

The answer is here, and Einstein had it. He always claimed that the existing quantum theory is incomplete. I propose that to complete the quantum theory you must link it with STEP theory. This will be expounded on further when I discuss the EPR debate.

The "firewall" resolution to the paradox. Some scientists suggest that the entanglement must somehow get immediately broken between the infalling particle and the outgoing particle. Breaking this entanglement would release inconceivable amounts of energy, thus creating a searing "black-hole firewall" at the black hole event horizon. This resolution requires a violation of Einstein's equivalence principle, which states that free-falling is indistinguishable from floating in empty space. This violation has been characterized as "outrageous"; one theorist has complained that "a firewall simply can't appear in empty space, any more than a brick wall can suddenly appear in an empty field and smack you in the face."

The complete quantum theory would take care of this and the firewall resolution would be a non-contender.

Non-firewall resolutions to the paradox.

Some scientists suggest that there is in fact no entanglement between the emitted particle and previous Hawking radiation. This resolution would require black hole information loss, an extremely controversial violation of unitarity.

This unitarity violation would not be allowed, when STEP synchronization is coupled with the complete quantum theory all information is available for every STEP, weather it is in the black hole or outside of the event horizon.

Others, such as Steve Giddings, suggest modifying quantum field theory so that entanglement would be gradually lost as the outgoing and infalling particles separate, resulting in a more gradual release of energy inside the black hole, and consequently no firewall.

Juan Maldacena and Leonard Susskind have suggested that the outgoing and infalling particles are somehow connected by wormholes, and therefore are not independent systems; however, as of 2013, this hypothesis is still a "work in progress".

The STEP theory in conjunction with the incomplete theory of quantum theory is my "work in progress" to resolve the paradox. Since all STEPs are synchronized they are also connected ant therefore not independent systems.

The fuzzball picture resolves the dilemma by replacing the 'no-hair' vacuum with a stringy quantum state, thus explicitly coupling any outgoing Hawking radiation with the information history of the black hole.

Stephen Hawking received widespread mainstream media coverage in January 2014 with an informal proposal to replace the event horizon of a black hole with an "apparent horizon" where infalling matter is suspended and then released; however, some scientists have expressed confusion about what precisely is being proposed and how the proposal would solve?

Einstein got it right about Special Relativity

Einstein's special relativity can be simplified by a metaphor where Einstein has a twin and together they unravel the mystery of time and distance from different points of view of observation.

Consider Einstein traveling at the speed of light of 186,000 miles a second (actually this would make him a photon), and his twin is watching him from 186,000 miles away.

Einstein's twin uses a stopwatch to time Einstein travel 186,000 miles. He found that it took two seconds on his stop watch to go that distance. This was because it took Einstein one second to travel 186,000 miles but it took another second for the light to travel from where Einstein was to where the retina of his twin's eyes were that saw the event.

As stated before Einstein's twin was one light second away (186,000 miles).

The time difference is referred to as "time dilution" in quantum theory. Also since Einstein's twin noted that it took two seconds to go 186,000 miles, he surmised that his twin only went 93.000 miles in one second instead of the 186,000 miles Einstein recorded from his perch on the light beam.

This shortening of distance is referred to by quantum physicist as "Lorentz contraction."

Einstein got it right about Warped Space

Before I discuss warped space-time, I would like to set some ground rules about the space-time web. I consider the web to be Fig#11.

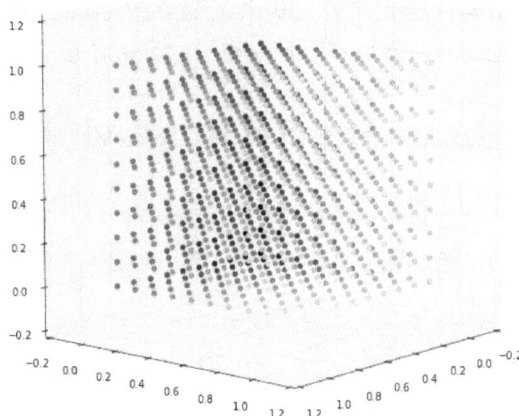

Fig# 11.A cubic entity

Space Time is cubic (see Fig#11 above) and not a space-time web with hyperspace around it like it is often depicted as seen below (Fig#12).

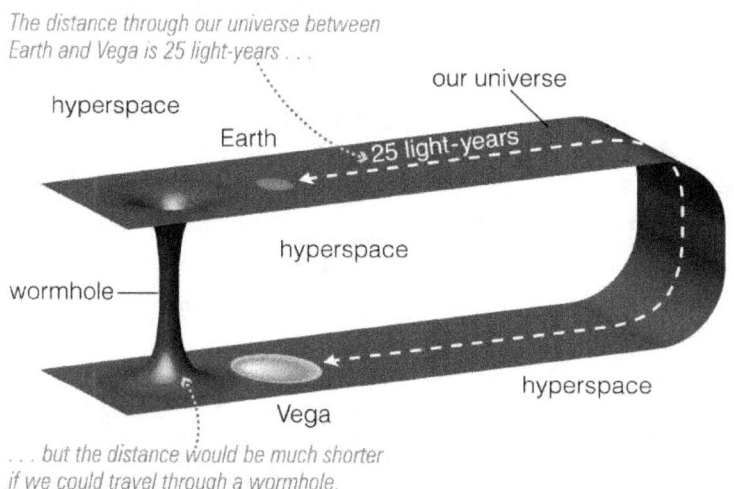

Fig#12. The universe depicted as a web in hyperspace.

Einstein's General theory of relativity suggest a warped space of geodesic geometry. Einstein showed that gravity is understood as a warping of the geometry of space-time, rather than as a pulling of objects on each other.

Objects move along geodesics lines (curved), which are determined by the warping of space-time.

If you were to travel straight from Key West to the Panama Canal by ship you'd probably think you were traveling in a straight line except for the bobbing of the ship. However, since the earth is curved, you would be traveling on a geodesic line (curved line). This is what Einstein means when he talks about warped space.

It's like the tesseract that's projected on a sphere but actually it's a cube.

Einstein goes on to show that space-time is further warped by matter especially if it has increased mass. The more mass, the more warping.

How can we understand this at the cosmos level as well as the Planck scale?

At the Cosmos level it has been shown many times over.

The most celebrated example was on May 29, 1919.

Sir Arthur Eddington led an expedition to test the new theory of gravity proposed in 1915 by Einstein's General Theory of Relativity.

It describes how any massive object, such as the Sun, creates gravity by bending space and time around it. Everything in that space is also bent: even rays of light. Consequently, distant light sources, behind the massive object, can appear in a different position or look brighter than they would otherwise.

In 1919, the Royal Astronomical Society (RAS) launched an expedition to the West African island of Principle, to observe a total solar eclipse and prove or disprove Einstein's General Theory of Relativity. They measured the deflection of the position of stars very close to the sun, an object that has the most gravity in the solar system. In order to measure the dim stars so close to the sun they used this solar eclipse that blocked out the sun's glare and made the dim stars more visible.

This first observational proof of General Relativity sent shock-waves through the scientific establishment, said Professor Ferreira. It changed the goalposts for physics.

To put this cosmos warped space-time in respective to a STEP web we need to look at the solar eclipse as described by Newton, Einstein and finely by the G.O.D. theory of everything (TOE). See Fig#13 below

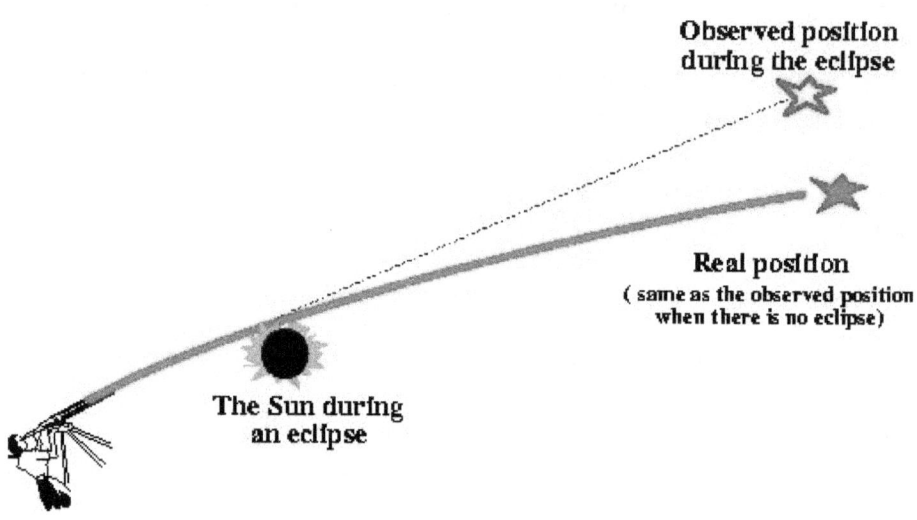

Fig#13. Solar Eclipse as depicted by NASA shows where the hidden open star would be by Newton's equations whereas Einstein predicted it to be behind the sun by geodesic geometry, which is shown here as a solid star.

In Fig#13 the sight paths seems to be a straight line for Newton's projection and a curve for Einstein's. However on closer inspection, of the computer graphic we see by digital representation that curve is actually a series of pixels that depict the curve.

See an expanded view of this in Fig#14.

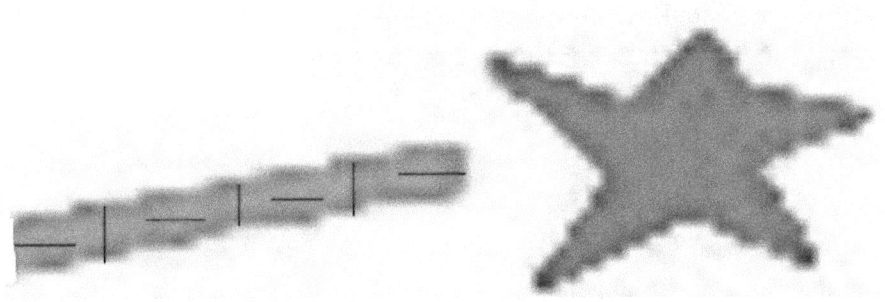

Fig#14. The seven STEPs leading to the star are shown here as distorted squares reveal the Higgs zero spin axis which when horizontal are passing through space-time, but when vertical are only in the time corridor.

When we look at space-time warping at the Planck's scale level we need to speculate on how it could be done.

STEPs that are distorted by localized string matter in the area would have a 50:50 over all ratio of cubes to cubes on a 3-spheres.

Therefore, any distortion or warping that takes place by this scenario is acceptable as long as the 50:50 ratio is maintained and the individual pair of space membranes stay parallel to each other.

When matter is moving through space-time the STEPs warp. The more concentrated matter is, the more the warping results.

The greater the mass the greater the warp. As individual STEPs warp in a region; "the hole is equal to the sum of its parts", so when matter is abundant, like the sun or even a black hole, then the warping is much more relevant, but the overall region of distortion is equivalent to the undistorted regions with respect to volume.

Space-time curves just as Einstein said, but gravitons pull like Newton said because the function of the gravitons are to pull all of the universe back to the original STEN and even to its final place of residence as a singularity in Eternity.

Another way to look at space-time warping is to consider the warping to be the synchronized rotational changes of the individual STEPs around string matter without the STEP being distorted. See Fig#15

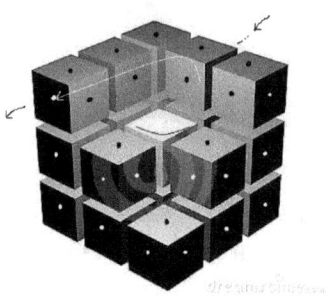

Fig#15. Dreamtime image to show space-time warping when the synchronized STEPs adjust to string matter. The small arrows represent a string boson (photon) warping around a string fermion (matter) in a block of ~20+ STEPs

The geometry of a STEP, along with its tesseract rotation, is engineered to allow a Planck particle strings to enter/exit in such a way that it can negotiate around other strings with ease.

By incorporation of STEPs into the geodesic path of travel, allows space-time to remain fixed while utilizing the tesseract rotation to afford string particles to step (curve) around others.

This modification of Einstein's General Relativity does not diminish it; it only enhances it. Only the description of how it operates needs to be addressed in the foreseeable future!

Einstein got it right about the duel nature of matter and antimatter being both a particle and a wave.

When I was younger, 1967-68, I had just spent two semesters at East Carolina University where I took two courses of calculus and two courses of Physics in preparation to become a chemistry teacher.

That summer I got a job with Hooker Chemical Company on Grand Island, N.Y. as a laboratory technician. It was second shift so it was quite quiet there and while doing GLC (Gas-liquid- chromatography) and x-ray analysis on flame retardant chemicals, I was able to read books and study chemistry, etc.

I never did go back to East Carolina University that fall but I did read a book about Einstein and the duel nature of particles and waves. In this book they said that if anyone can explain this duality they would surely get the Noble Prize.

I thought about this often, I could not get this off my mind, I pictured soap bubbles, strings, loops, etc. but I couldn't come up with a logical way that this could be explained.

Well, last night November 11, 2013, I had an epiphany and found myself thinking about this again, only this time I had an answer.

What I'm about to explain might be already documented or inferred and I missed it, but here is what I came up with:

Normal matter and antimatter are considered to be one dimensional strings that are 1 Planck's length long.

A particle matter, say a neutron for example is, said to have a diameter of ~10^{-15} meters which results in a volume of ~10^{-45} meters cubed. A single Planck string occupies a volume of 10^{-105} meters cubed.

Since the volume of a STEP tesseract on a three sphere is ~10^{-105}, we can calculate the number of STEPs that the Planck string occupies by utilizing this volume as a standing wave, which is $10^{=45}/10^{-105} = 10^{60}$ STEPs.

This volume of 10^{60} STEPs might be thought of as an electron Shroud.

A single strand of matter will not only operate in a precise confined volume as a standing wave, but it will also be a string particle. End of story!

Part IV

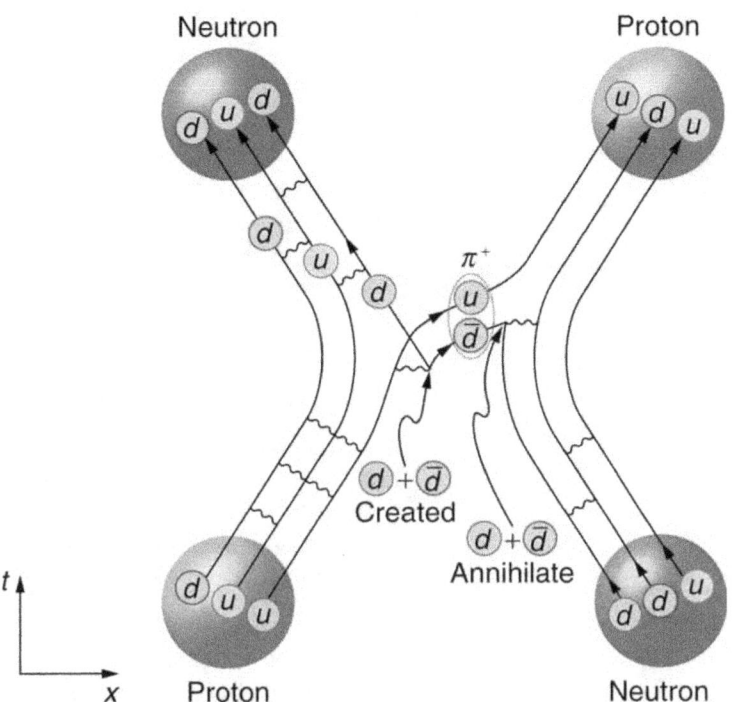

After Einstein

"The most incomprehensible thing about the world is that it is comprehensible."

"We can't solve problems by using the same kind of thinking we used when we created them."

"The whole of science is nothing more than a refinement of everyday thinking."

Big Bang and Inflation

My idea about the big bang and inflation is slightly different than other proposed theories but still in the realm of scientific possibilities.

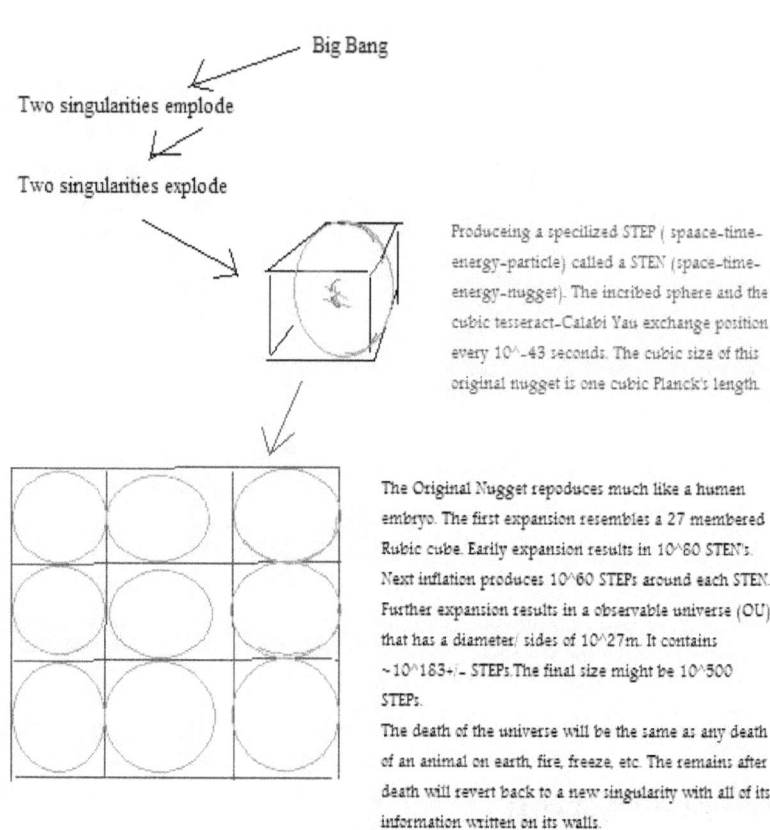

Big Bang

Two singularities emplode

Two singularities explode

Produceing a specilized STEP (spaace-time-energy-particle) called a STEN (space-time-energy-nugget). The incribed sphere and the cubic tesseract-Calabi Yau exchange position every 10^-43 seconds. The cubic size of this original nugget is one cubic Planck's length.

The Original Nugget repoduces much like a human embryo. The first expansion resembles a 27 membered Rubic cube. Earlly expansion results in 10^60 STEN's. Next inflation produces 10^60 STEPs around each STEN. Further expansion results in a observable universe (OU) that has a diameter/ sides of 10^27m. It contains ~10^183+/- STEPs.The final size might be 10^500 STEPs.

The death of the universe will be the same as any death of an animal on earth, fire, freeze, etc. The remains after death will revert back to a new singularity with all of its information written on its walls.

Fig#16. Big bang as proposed by the G.O.D.

The Big Bang and the congealing of 4.76 % of its dark energy into the particle zoo only took~ 4 Planck tics of time to create in the universe today. Also, during the very first tic of time the graviton was spun off from the Eternal force of the universe.

The Big Bang was caused by the entanglement of two singularities that resulted in an extremely hot Space-Time-Energy-Nugget (STEN) housing Zero-Point Energy (ZPE) with in. This happened during the first Planck's tic of time. This first tic also enabled an onion-skin path way to develop between the two dark energies produced from the 2 singularities. The two dark energies (De) resemble tesseracts and exchange location in the STEN every 10^{-43} sec

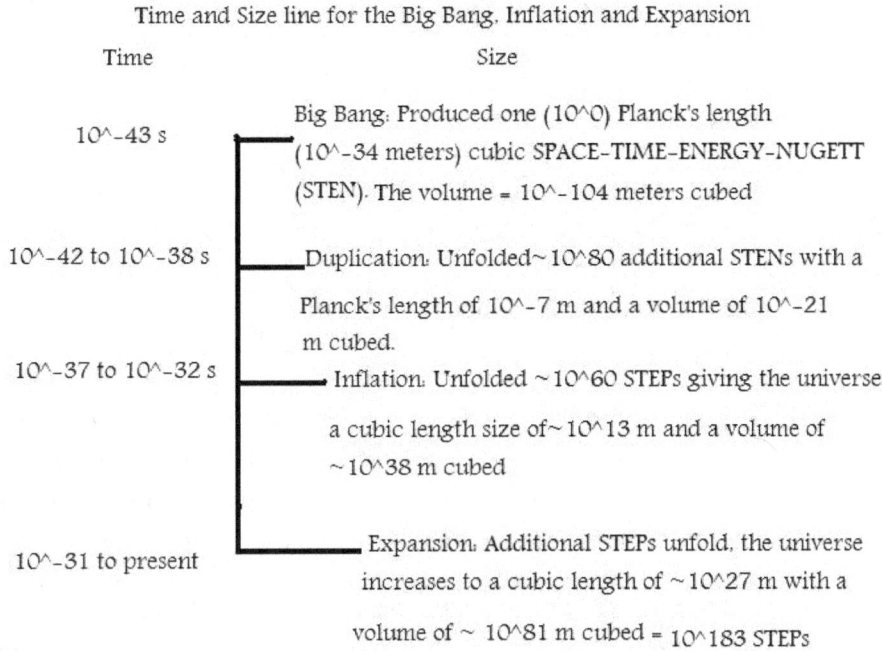

Time and Size line for the Big Bang, Inflation and Expansion

Time — Size

$10^{\wedge}-43$ s — Big Bang: Produced one ($10^{\wedge}0$) Planck's length ($10^{\wedge}-34$ meters) cubic SPACE-TIME-ENERGY-NUGETT (STEN). The volume = $10^{\wedge}-104$ meters cubed

$10^{\wedge}-42$ to $10^{\wedge}-38$ s — Duplication: Unfolded~$10^{\wedge}80$ additional STENs with a Planck's length of $10^{\wedge}-7$ m and a volume of $10^{\wedge}-21$ m cubed.

$10^{\wedge}-37$ to $10^{\wedge}-32$ s — Inflation: Unfolded ~$10^{\wedge}60$ STEPs giving the universe a cubic length size of~$10^{\wedge}13$ m and a volume of ~$10^{\wedge}38$ m cubed

$10^{\wedge}-31$ to present — Expansion: Additional STEPs unfold, the universe increases to a cubic length of ~$10^{\wedge}27$ m with a volume of ~ $10^{\wedge}81$ m cubed = $10^{\wedge}183$ STEPs

Table#2 The 10^{80} STENs is based upon Eddington's 1939 estimate of proton/neutrons in the universe. The 10^{60} STEP unfolding during inflation is based on the volume needed to create a shroud around each STEN. The additional STEPs make up the remaining OU

The onion skin appeared as a result of a small portion of a tesseract to coalesce into the thin onion skin pathway and produced a congealed particle.

The particle that was formed was an "O" ring-like graviton which navigated throughout the onion skin corridor.

The graviton had the potential to pull the STEN back to its origin upon cooling. The STEN was the size of a STEP and in fact is a specialized STEP.

But before that could happen, a second tic took place along with several more, resulting in further congealing and all of the elementary particles of the universe were formed. The STEN was what L. Garrett refers to as an "exceptional simple lie group".

It has all 248 elementary particles along with their ghost shadows. (Note: Garrett does not include these ghosts, and I think he still refers the particles as points where here I consider them Planck strings with supper-symmetry shadows).

During this time period the STEN was also duplicating its self many times until there were ~10^{80} such nuggets, each the size of a STEP and all containing the "exceptional simple lie group". The volume of these 10^{80} STENs was 10^{-21} meters^3, this is about the size of a HIV virus. Its cubic length is 10^{-7} meters.

Note: The Big Bang might be looked at as a metaphor to the conception of a new born baby in that the two singularities are that of a sperm and an egg. When the sperm entangles with the egg it produces a proliferation of cells that have all the information it needs to enlarge and function.

On the next 4 tics (tic 10^{-36} through 10^{-32}) all of the 10^{80} STENs created their own individual shrouds by releasing enough STEPs to give them a cubic length of 10^{13} meters (which is the ~ size of the Kuiper belt). See **http:htwins.net/scale2**

From Wikipedia:*" The **Big Bang** theory is the prevailing cosmological model for the universe from the earliest known periods through its subsequent large-scale evolution. It states that the universe was in a very high density state and then expanded. If the known laws of physics are extrapolated beyond where they are valid there is a singularity. Modern measurements place this moment at approximately 13.8 billion years ago, which is thus considered the age of the universe. After the initial expansion, the universe cooled sufficiently to allow the formation of subatomic particles, and later simple atoms. Giant clouds of these primordial elements later coalesced through gravity to form stars and galaxies."* **My main alteration of this theory is that it took two singularities for creation and also that the first**

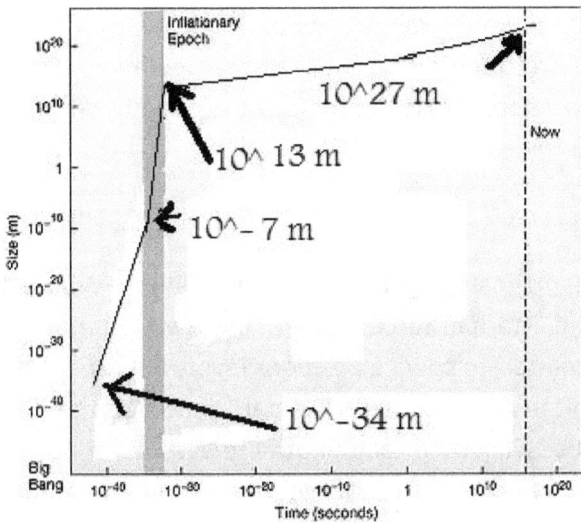

Fig#17. All 10^{80} STENs were formed in the first 7 tics of time and was followed by rapid inflation (from 10^{36} t0 10^{32} sec,) in the next 4 tics. Each STEN became surrounded with 10^{60} STEPs as shrouds. Further expansion followed until our OU enlarged to 10^{27} m.

elementary particles were unified as L. Garrett's "high dimensional objects" or "exceptional simple lie groups" and where rearranged, separated, recombined etc. after inflation.

Collectively the 10^{80} "exceptional simple lie group" with their shrouds had a volume size of $\sim 10^{38}$ meters^3. See Table#2.

Currently the observable universe (OU) stretches in all directions from our vantage point and is filled with Space-time energy nuggets (STEN) and Space time energy particles (STEPs). The number of STENs and STEPs to date relate to $\sim 10^{183}$. Of course space-time is continually expanding and this number is expanding accordingly.

The strong force came into existence when inflation occurred, and shortly thereafter the electromagnetic force (EMF) and the weak force appeared.

The Five forces of the Universe

The universe has five forces, not four. The four forces of gravity, strong and weak force plus the (EMF) are often discussed with the operation of our universe, but actually a super force, that the other 4 forces spun off of, is still in play.

I call this force "The Eternal Force".

The ZPE/ZPM housed in the original nugget as well as all STEPs is regulated by this Eternal force. The Eternal force causes the two alternating tesseracts of dark energy (matter) to change positions (every 10^{-43} sec.) and also rotate when synchronization is needed. The dark aspect of the two Calabi Yau/ tesseracts appose each other and neutralize each other out so they are not felt in our everyday life like normal matter and energy do. Only gravity feels the effect of this dark energy when it is concentrated as dark matter).

However, at the first tic of time when the diameter of the tesseract cube on a 3-sphere was greater than the sides of the walls of the other cube, the Eternal force was felt as pressure on the space membrane walls of the original nugget. During this first tic of time, some of the pressure was elevated by enabling an onion-skin to develop between the two tesseracts, of the ZPE, and as stated before it congealed some of the ZPE into a Graviton which has the potential to pull the STEN back to its origin, upon cooling. (Note: this is the first "separation of forces").

The pressure was further reduced by concentrating the ZPE into ZPM. This resulted in all 10^{80} STENs to contain dark matter and in turn the shroud STEPs, 10^{60} (that were produced during inflation around each STEN) to have ZPE by the reverse process.

The 2nd tic through 10^{-36} sec. resulted in further pressure being reduced by producing element particles of the universe to congeal out of the tesseract on a three sphere. These elementary particles left wormholes in the 3-sphere which allow the two tesseracts to have equal volume and equal sides/diameter.

The conversion of ZPE to ZPM follows an Einstein-like equation $De=DmC^2$. The particles emerged as neutron STENs, 10^{80} of them, each with 248 elementary particles (strings) compartmentalized as what L. Garrett calls "the high dimensional object".

The strong force emerged to keep all the 248 elementary particles in their respective wormholes (2nd separation of the Eternal force).

However, inflation took over. Inflation resulted from leaking of ZPM from the 10^{80} STEN's. This leaking was by the conversion by dark matter back to dark energy. This dark energy manifested its self into STEPs that differ from the STEN by having alternating dark energy and lacking any elementary particles. It does have all 11 dimensions so elementary particles can move in and out of them freely. The weak and the electromagnetic force (EM) separated out and the Higgs field was activated at this time between 10^{-32} and 10^{-12} sec.

Also, after inflation occurred the 248 elementary particles "moved out. "

They moved into the shroud STEPs surrounding the STENs. The STEN became the "scaffolding" or "fractal tree" for galaxies to form after the particle zoo reorganized. More will be said about this when we discuss continuous Dm, De conversion.

In the later years of Einstein's life, he spent many tedious hours toiling over equations to unify the electromagnetic force (Em) to that of gravity.

Here I postulate that not only these two forces are related but that all four forces that are currently discussed are derived from a single force--- The Eternal force--- the force that drives the universal synchronized steps which in turn are governed by choices created by mind and matter.

Particle Zoo

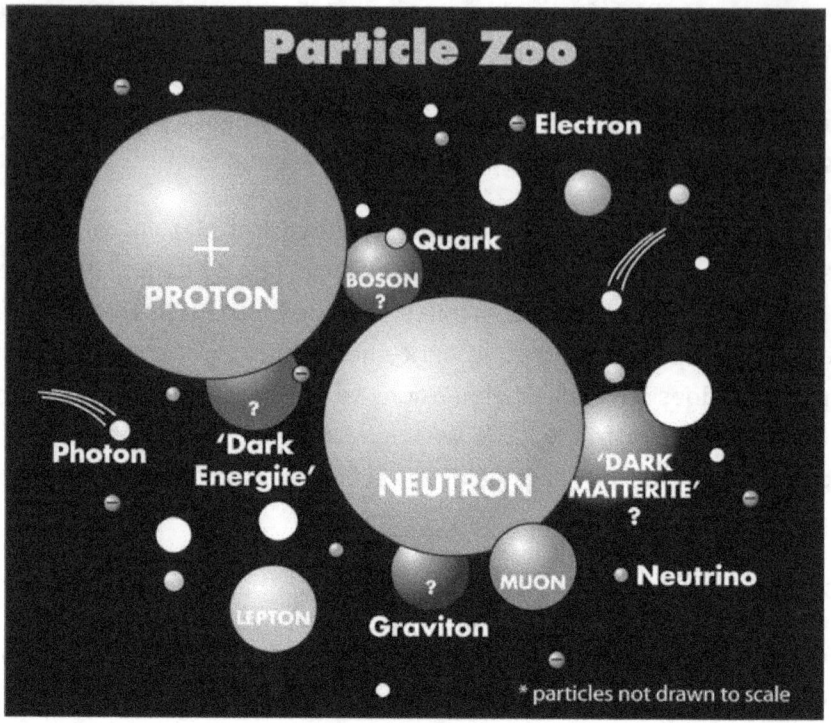

Fig#18. This Particle Zoo image, reproduced from Bing's WWW is interesting, since I would have the Dark Energite' and Dark Matterite' being examples of a STEP and a STEN (specialized STEP) respectively.

The 248 elementary particles acting in concert as "high dimensional objects" not only moved into the shroud STEPs, but also into the newly formed STEPs that were leaking out from the STENs, but at a much slower pace than during inflation.

These "high dimensional objects", all 10^{80} of them that came from the 10^{80} STENs were some-what unstable. Two down quarks in them become activated along with an up quark. As a result of this, 10^{80} neutrons formed.

These neutrons were highly active at the current temperature of ~ 17 degrees K. They moved at near light speed and often they would smash into each other, causing some of them to silence an up quark and at the same time activate a down quark. Also, an electron and an anti-neutrino escaped from the neutron plus a measured amount of normal energy.

This converted the neutron into a proton and caused a spontaneous symmetry brake which resulted in the electroweak epoch to split into the EM and nuclear weak force. This is known as Beta decay and is depicted as a Feynman drawing. See chapter IV heading image.

Note: The neutron as stated above is different than current thinking. It has all 248 elementary particles plus the Sparticles which are inactive except for quarks and gluons. A similar situation applies for protons except for the beta decay loss.

Eleven Dimensions

Earlier, I spoke of a sphere inscribed in a cube having equal volume's both contained in a Planck's cube (STEP). When you inscribe a solid sphere in a cube, the volume that is occupied by the sphere is 52.38 % and the remaining volume of 47.62 % is left to fill the cube.

However, if the sphere is reduced to 47.62 % volume by shaving it ever so slightly and drilling cork screws, and other worm holes through or inside it, then the sphere and the cube will have identical volumes.

The shaved outer surface of the sphere and drill wormholes have a volume of 4.76 %.

Now consider the sphere and the cube to be dark matters (Dm) apposing each other which gives net zero-point matter (ZPM). The shaving and the hole drilling was really the process of 4.76% of the dark matter in the sphere congealing to normal matter (Nm) string particles by way of;

$$Dm= [De - Nm]/C^2.$$

These left pathways for the normal matter to flow. In this case the Nm includes both force particles and matter particles.

These two ZPM's trade places every Planck tic~10^{-43} sec. by way of tesseract rotation. The sphere (cube on a 3-sphere) always will have 6 wormholes which make it a Calabi-Yau manifold (see Fig# 14). This gives a STEP 10 dimensions and the shaving of the sphere created the onion ring which is the 11 dimension.

The shavings is a graviton.

This condensed dark matter/energy that left wormhole path-ways throughout the Calabi-Yau tesseract on a three sphere resulted in the particle zoo.

Fig#20. According to Lisi Garrett there are 226 known elementary particles and an additional 22 more totaling 248.

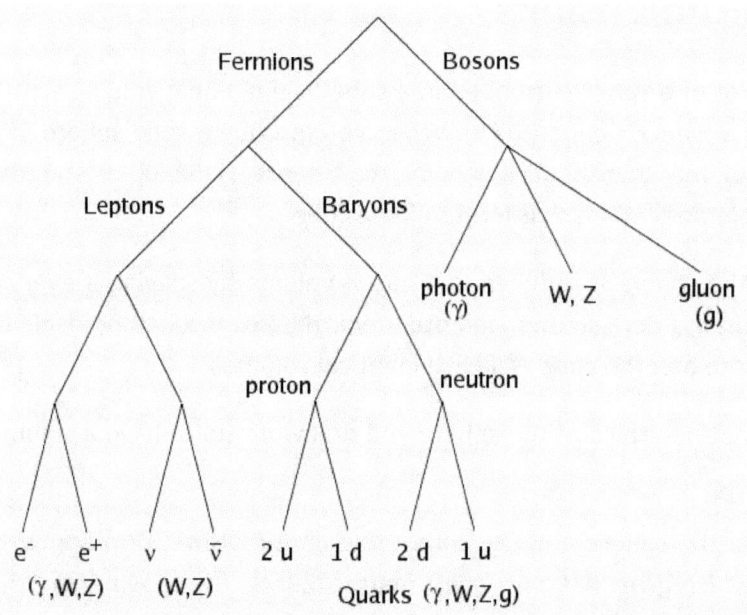

Fig#19. ©2014, Kenneth R. Koehler. Sparticles are not shown here. They are the super-symmetry particles. Garrett's 248 elementary particles include the antiparticles and the graviton but not the super-symmetry ones. If he did there would be an additional 248 more. The lower rows of each group show some of the elementary particles.

108

Creation of the particle zoo.

The particle Zoo is Garrett's 248 elementary normal particles (see Fig#20) and the additional 248 super-symmetry ghost (hidden) particles that are referred to as Sparticles.

The graviton force particle operates in the onion skin only as a two spin boson.

The Higgs particle moves through the axillary worm hole which passes through the center of the Calabi-Yau projected on a three sphere as a zero spin.

The four other worm holes are reserved for the remaining four fermions and boson particles such as the +1/2, -1/2, +1, -1 spins respectively.

The fifth 3/2 spin worm hole is where the non-detectable super-symmetry Sparticle shadows habitat.

This congealed dark matter into normal make up the particle Zoo.

According to Dr. Anthony L. Garrett this zoo is a dimensional E8 objects (see Fig#20).

 Each particle is considered a point particle by Garrett, but since a point particle cannot be detected unless it is at least a Planck's length, we need to set the particle as a one Planck's length string.

 Currently, the total number of known particles is a handful less than 248 but the search is on to discover more particles. Also as noted before, the Sparticles double the number.

Fig#20. This is Garrett's E-8 lie group that has 248 elementary particles with some important interactions, Creative Commons Attribution-Share Alike 3.0 Unsupported license.

Spin

Spin is an interesting concept with respect to the quantum theory. It was basically developed during and after Einstein's era.

Quantum spin should not be confused with the classical spin of a top. It relates (among other things) to the electrons in which Paul's exclusion principle require electron pairs to have opposite spins to be in an orbital of an atom.

Earlier I noted where specific corkscrew worm holes resided and where some of the matter particles can be located. It is these corkscrew wormhole dimensions of space that give particles their "quantum spin".

The boson force matter particles all have a +1 spin and travel in the+1 right hand turn of the cork screw wormhole.

Their counterpart, the anti-force string particle, have a -1 left and travel in the -1 left hand turn of the cork screw wormhole.

Fermion matter, such as the electron, have a - 1/2 right hand spin and travel in the -1/2 right hand turn of the cork screw wormhole.

Their counterpart, the Fermion antimatter, such as a positron, have a +1/2 left hand turn and travel in the +1/2 right hand turn of the cork screw wormhole. .

Note: The Pauli's exclusion principle requirement that electron pairs must have opposite spins to be in an orbital of an atom is valid under these space-time dimension parameters. The Planck length string matter particles simply flow through space-time with and against the twist of the wormholes. See Fig#22.

Fig#21-23. Concepts of spin.

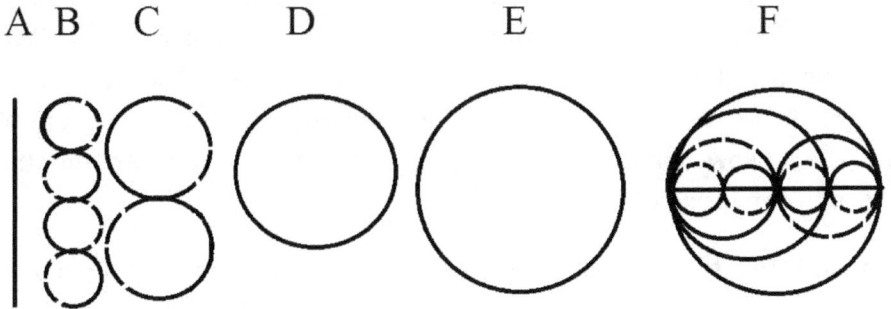

A B C   D   E   F

Fig#21. These 2 dimensional images depict the 5th through 11th worm holes/onion skin conduits that normal string matter/ antimatter (Fermions), force string particles (Bosons) and Sparticles transverse through space-time. A through E highlight the geometry of the conduits along with the particular particle in question. The descriptions are as follows:

Fig#21. A is the 5th dimension that is an axillary worm-hole that the Higgs Boson travels through. The Higgs Boson has a zero "0" spin and can be considered to be found in the center of the cubic 3-sphere. Therefore, it has a 0 radius away from its self (this sounds a little stupid but it will become clearer as you read on).

B is the 6th and 7th dimensions that are two corkscrew worm-holes which are a distance of ½ radius length away from the axillary worm-hole. The normal string particles (Fermions) move through these pathways ether as matter in the solid curves or as anti-matter in the broken curves

C is the 8th and 9th dimensions that are two corkscrew worm-holes which are at a distance of 1 radius length away from the axillary worm hole. The force string Bosons travel through these conduits ether as matter interacting Bosons (solid curve) or as anti-matter {if any} interacting Boson (broken curves).

D is 10th dimension and is where the shadow Sparticles reside. They are 3 radius length away from the axillary worm-hole which results in a 3/2 ring-like worm-hole that is not 1 Planck unit long in diameter, so they cannot be detected in our universe! Sparticles move in the 3/2 spin twisted worm hole. They are hidden from our universe by being totally in the STEP universe that lack a one Planck length to be detected. The Sparticle is important because it is entangled to its counterpart. It is also important because it allows for super symmetry and the M theory to incorporate the eleventh dimension of gravity.

E is the 11th dimension and is where the Graviton operates. It is 4 radiuses away from the axillary worm-hole which gives it a 4/2=2 spin connotation which is the exact size needed to produce an onion skin around the 3-sphere. This allows the 2-spin graviton to operate in the onion skin. The Graviton moves in and out of our universe as depicted in Fig#10.

F is the 2-dimensional representation of the 5th through 11 dimensions that are proposed by Dr. Witten in his M string theory. For the complete 11 dimension shown as a cube with the 3-sphere inscribed within, see Fig#02B and Fig#5B.

Like the statically quantum analysis for the shells (95 % +/-) of electrons to roam around the nucleus of an atom, the spin of a string is also a probability, so when I refer to a +1/2 twist radius of a corkscrew, it too has a probability value that the spin of a string will be located there.

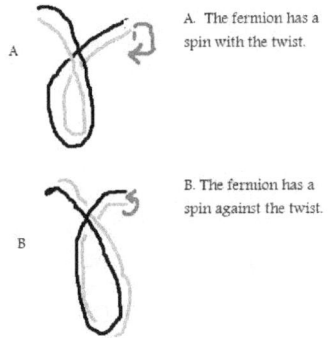

A. The fermion has a spin with the twist.

B. The fermion has a spin against the twist.

Fig#22. Wolfgang Pauli's opposite spin

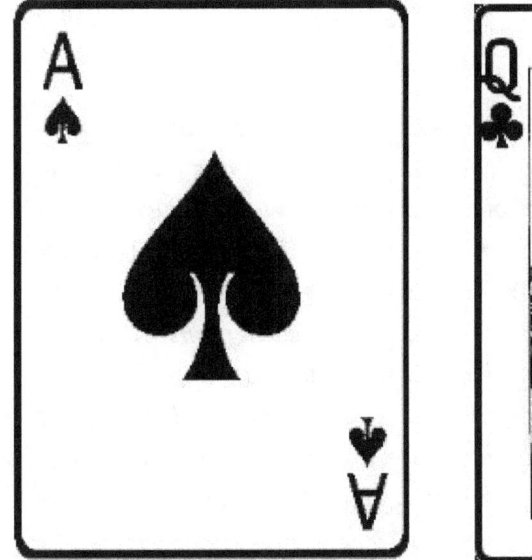 

Fig#23. Stephen W. Hawking images that explain spin.

http://www.markusehrenfried.de/science/physics/ hermes/whatisspin.html

There is another way to look at spin: it tells us something about the symmetry a particle. Stephen W. Hawking explains this aspect of spin in chapter 5 of his book A Brief History of Time with a nice example: (see Fig#23)

"A particle with spin zero behaves like a point: it looks the same way regardless of the direction from which you look at it.

Something which has a symmetry like the playing card on the left needs a full rotation of 360° until it looks again the same. That is the sort of symmetry a spin-1 particle has: after a full rotation it is again in the same state. (Obviously it makes no sense to say something like 'a particle looks the same' as we are not able to see it, but particles have well defined states which can be detected, that's why we can say that a particle is again in the same state).

A spin-2 particle behaves under rotation like the playing card on the right hand side. It already looks the same (is again in the same state) after half a rotation (180°).

The electron is a spin-1/2 particle, and now things become strange: a spin-1/2 particle needs two full rotations (2 x 360°=720°) until it is again in the same state. There is nothing in our macroscopic world which has a symmetry like that. Common sense tells us that something like that cannot exist, that it simply is impossible. Yet that's how it is. Actually, it is even relatively easy to set up an experiment in a lab which demonstrates that electrons behave exactly in this weird way: if you 'turn' them around once they are not in the same state but in minus that state and only after another full rotation they are again in the state they had initially. There is no way to explain this if we imagine spin as a little arrow in the three-dimensional space of our everyday life"

With the knowledge that space and time can be mapped with~10^{183} STEPs it is possible to use classic examples to explain Spin.

Einstein used his creative energy to change the way we see the universe today. He often used "thought experiments" to develop his unique ideas about time, space and gravity. Sometimes he would mentally visualize himself traveling at the speed of light, or being in an elevator in a free fall (gravity independent).

I used similar metaphors when I thought about space and time. I first wanted to make a cube and a sphere to have a common dimension. The diameter of the sphere and the sides of the cube are unequal because I saw space as curved but others considered the universe to be flat.

The problem was that an inscribed sphere in a cube has more volume, 4.76% more, than the remaining volume left in the cube.

I mentally saw myself inside a spherical apple, carving out worm holes in it so that the volume is the same as the cube, and each would have similar dimensions of diameter and sides and volume.

Being a food research scientist, I saw the slim region between the spherical apple and its cubic crate to be implied as a thin onion skin.

Six worm holes, one onion skin and two membranes plus the speed of light (length/second) gave me the eleven dimensions I needed to satisfy the requirements of M theory's dimensions. The requirements for our universal dimensions can know be looked at as real dimensions of space time instead of some of them being only mathematical entities generated from just theoretical equations.

That is why I use the twists to be radius distances from a common axil of 0 spin to solve the dilemma of spin.

Stephen Hawking presents a strong case but could not explain -1/2 spin with ordinary examples.

With his outlook as an atheist he does not have the advantage of Deity insight.

If he embraces the "Grand Original Design" concept he probably could return to the status, he once enjoyed when he was open an Eternal view.

Creation trumps spontaneity.

Marriage through gravity incorporating STEPs

String theory and Einstein's loop quantum gravity are both trying to achieve the same goal - quantified gravity and unify General Relativity and quantum mechanics into a seamless theory of everything.

However, they are doing it from opposite directions - string theory starts in the realm of quantum mechanics and tries to introduce gravity as just another string (the graviton).

Loop Quantum Gravity tries to quantify gravity (and thus space- time) using General Relativity as a starting point, and presumably hopes to try to introduce the features of quantum mechanics.

This is a good starting place to merge both theories together.

By considering the quantum theory to be incomplete, as Einstein, Podolsky and Rosen (EPR paradox) did, and then by incorporating the STEP concept, the theory can be considered complete and the unification of the theories can solve the gravity issue at the large level of the universe as well as the small.

The closed loop of the graviton has a diameter of one Planck's length when in the quantum region of space-time that is found between the two tesseract of space-time-energy where the onion ring volume is found. This is true only when the graviton is at its maximum: see Fig#21. When not at the maximum it is hidden from the universe.

Gravity whether its micro or macro operate by one simple mechanism. The graviton travels at the speed of light and has a single purpose of pulling all matter-antimatter back to its original site, (the original singularities) as discussed earlier.

The concept of a graviton being hidden supports the parallel theory of Lisa Randall where she claims that the graviton spends more time in a parallel world than in ours and therefore making it much weaker than it seems.

The STEP can be looked at as a parallel world since its interior is hidden from ours. When the O-ring graviton is not at the equator of the onion skin then it is hidden in the other parallel world.

Although Einstein only dealt with 4 dimensions his introduction of time as the 4th dimension, opened the door for the advanced super-string theory of 11 dimensions called the M theory.

In 1994, a string theorist named Edward Witten and other researchers considered the five different versions of string theory might be describing the same thing seen from different perspectives.

Witten proposed a unifying theory called "M-theory", in which the "M" is not specifically defined, but is generally understood to stand for "membrane". The words "matrix", "mother", "monster", "mystery", "magic" have also been claimed. M-theory brought all of the string theories together. It did this by asserting that strings are really 1-dimensional slices of a 2-dimensional membrane vibrating in 11-dimensional space.

Witten added the 11 dimension to the string theory and it was in fact related to gravity.

The loop theory already called for 11 dimensions so this was a plus to unite the two theories.

Now since there are ~10^{183} STEPs in our Observable Universe and of them there are the original 10^{80} STENs each providing an active graviton, it is logical to infer that the large: operates as the sum of its small parts and that the gravity acting in the bulk is the same as gravity on the Planck's scale.

Also the theory of everything is satisfied because the necessary Sparticles that are needed to satisfy the M theory and super symmetry are found entangled with the particle zoo.

Lastly, the great minds still living that followed Einstein in time such as Anton Zelinger, Leonard Susskind, Anthony Lisi Garrett, Lee Smolin, Stephen Hawking, Lisa Randall, Edward Witten etc. are only almost right because they have not quite closed the circle on the entanglement of Quantum-mechanics and General-Relativity and STEP.

Einstein brought together space with time, quantum packet photons with waves and postulated that Quantum theory was incomplete.

My hope is that the current theoretical physicist can gain some inspiration from my thesis between the covers of this book and by use of quantum theory entangled with relativity they can follow Einstein's vision.

Using their creativity and their knowledge they will embrace Einstein as never before.

"Imagination is more Important than Knowledge"

Creativity is spawned by the entanglement of Imagination with acquired Knowledge.

Creativity is more important than knowledge but knowledge includes knowing that knowledge enhances creativity.

The Incomplete Quantum theory

Einstein became embroiled in a hotly disputed discussion on the Idea of whether or not particles can communicate with each other at a distance without ever coming in contact with each other.

His premise was that information cannot travel faster than the speed of light based upon his special theory of relativity.

Einstein and his associates Podolsky and Rosen (EPR paradox) preferred to think that this "Spooky action at a distance" of two particles, i.e. two electrons, must be due to an unknown factor hidden in the jowls of space-time.

With the advent of a granular universe of ~10^{183} STEPs, which are all synchronized with each other, it is plausibly that these two electrons particles communicate by the synchronization of space time so that what is known in one part of the universe is instantly know throughout the universe.

This would fit with what Einstein felt was particle entanglement.

This does not violate Einstein's velocity maximum principle!

If you think about a wind up clock with all its gears that move in harmony as time passes, you become aware that when one gear moves in one location all the others move at the same time but at different ratios. This of course is assuming that this clock is friction free with perfect mechanical tolerance. Now, think of the universe as a clock, when one STEP is disturbed by a particle all of the other 10^{183} STEPs are instantly informed by the synchronizing of the STEPs and not by the speed that the particle travels through it.

Thus this does not violate Einstein's terminal velocity requirement of matter, but allows communication to be possible between two distant particles.

The STEPs are Einstein's hidden factor which make "Spooky action at a distance" possible.

Einstein was right by pointing out that the quantum theory is incomplete because currently, or at least then, the theory did not completely engulf the concepts of super string theory or its implication with respect to Space-Time-Energy granularity.

Universal Clock

When we think of the universe we most likely think of it at its current size being on the order of 10^{27} meters cubed for the observable universe. But since we just discussed the big bang and the inflation of size that the universe went through, it is not hard for me to think of it when it was the size of a grapefruit or better yet, the size of an old fashion windup alarm clock I used when I was in High School. My clock had a round face with two hands attached to an axle and gears inside the frame that it sat in.

119

The wind up clock worked on the principle that the stored energy from the tightened coiled spring would uniformly drive the gears at acceptable ratios as to allow the hour hand and minute hand to show when it was time to get out of bed.

One major feature of this clock is that every gear in the clock is synchronized with each other to produce the desired result. If all the mechanisms of the clock were friction free and the spring could release a uniform amount of energy, then it would be possible to claim that the clock is a closed system where every part knows what every other is doing and act accordingly at an exact precise way.

Now, if we consider our universe to operate the same way as the clock, we could say that every STEP would know what every other STEP was doing at all times. This would be true if the Universe was the size of the clock or if it is the size of the current universe which as stated earlier has ~10^{183} space-time-energy-particles.

This concept might bring you to wonder if information can transform across the universe faster than the speed of light.

The answer is yes and no!

Information can only move through STEPs at the speed of light. But like the gears on a clock when one STEP moves the entire system reacts to the movement (Note; STEPS tic every 10^{-43} sec.) and thus, what happens in one part of the universe is felt in another part by the synchronization of the STEPs and not by the transformation of information through it.

This is the answer to Einstein's "Spooky action at a distance phenomenon".

Einstein hotly contested the notion "that information could travel through space-time faster than the speed of light." He was absolutely right and he chose to explain it in terms of hidden variables associated with all matter.

These hidden variables, associated with matter, have been thoroughly disputed by Bell's theorems and the quantum theory advances of Heisenberg's' uncertainty principle.

But if you consider the hidden variable to be in the STEP instead of the matter-antimatter in motion, then Einstein's "Spooky action at a distance" does not violate Bells theorems nor does it violate his relative theory where the speed of light is of concern. The quantum entanglement of matter-antimatter particles etc. are allowed by this interaction of STEPs.

Questions of causality and locality are of no longer a concern since the "canvas instead of the paint" is the driver of Spooky actions.

Ultimate fate of the Universe

The several theories about the end of our universe such as "Big Freeze", "Big Crunch", "Big Rip", "Big Bounce", "Never ending Multiverse" or the "False Vacuum, Multiverse", and "Phase change" are some of the proposed theories of how our universe will end! **http://en.wikipedia org/wiki/Ultimate fate_ of the _universe.**

Of them, I think the "Big Bounce" is the most realistic, because as dark matter leaks out of STENs and produces additional STEPs with dark energy the universe become less self-sustaining.

When the universe sheds all of its dark matter into dark energy the total expansion will reach a finite conclusion {10^{500} seconds}. This allows the graviton to get the upper hand and bring the entire expanded universe back to a single nugget which can reduce to a singularity (or two) and lodge in eternity of which it came. All information of the universe's history will be stored on or in the singularity(s). (Note: singularities are much smaller than Planck's volume   so any speculation on this would be meaning less).

I call this the "Universal Library Collection."

From Fig#06A we see that it took 13.7 billion years to dilute or consume ~ 1/3 of dark matter.

At the end of this expansion the universe may follow some of the proposed theories mentioned earlier, but "information can never lost".

Gravitons will pull matter into individual black holes and in turn unite these black holes until there is only one (or two) black hole(s) remaining. This black hole(s) is further reduced to a Planck size nugget(s) containing all the informational history of the universe. No radiation or information will be lost, the information will be inscribed on the onion skins of the nugget.

The nugget(s) will then be reduced to a singularity(s) that is (are) infinitely smaller than the nugget(s) which is (are) too small to speculate as to what happens next.

Fig#06A. NASA info. Indicates that the early universe did not have any dark energy, but as the universe expands and ages, it developed ~69% dark energy, and the dark matter diluted or reduced to ~27%.13.7 yrs. ago dark matter was 63% of the universe. This is a similar % just after inflation (10^{80} $+10^{60} = 10^{60}$ & 80/140 = 57 %) see table # 2. The 6 % difference is rounding errors, poor estimates of the shroud STEPs and/or expansion of the OU to 380,000 years old.

The End of the Universe (in review**)**

The end of the universe is analogues to the end of life of an animal, such as homo-sapiens. The end of the universe has many theories such as:

 Big Freeze - Universe expands until it freezes out.

 Big Rip - The universe rips apart upon expansion.

Big crunch - Universe stops expanding and then contracts back to where it came and in-turn produces another bang.

Phase change - Universe acts like water and causes a phase change -like going from a liquid to a solid.

I hypothesize that the Grand Original Design is complete when all of the information that is gathered during its existence will be returned and be stored in one or two singularities that are part of the infinite Eternity....

"Universal Library Collection"

This concept is supported by the incomplete quantum theory that states "information can never be lost or destroyed" and I see no reason to think that the STEP theory would alter this.

This concept would incorporate much of the big crunch but lack the production of another big bang.

From Wikipedia, the free encyclopedia

In physical cosmology, the **Big Crunch** is one possible scenario for the ultimate fate of the universe, in which the metric expansion of space eventually reverses and the universe recollapses, ultimately ending as a black hole singularity or causing a reformation of the universe starting with another big bang. Sudden singularities and crunch or rip singularities at late times occur only for hypothetical matter with implausible physical properties.

Overview

If the universe's expansion speed does not exceed the escape velocity, then the mutual gravitational attraction of all its matter will eventually cause it to contract. If entropy continues to increase in the contracting phase (see Ergodic hypothesis), the contraction would appear very different from the time reversal of the expansion. While the early universe was highly uniform, a contracting universe would become increasingly clumped. Eventually all matter would collapse into black holes, which would then coalesce producing a unified black hole or Big Crunch singularity.

The Hubble Constant measures the current state of expansion in the universe, and the strength of the gravitational force depends on the density and pressure of matter in the universe, or in other words, the critical density of the universe. If the density of the universe is greater than the critical density, then the strength of the gravitational force will stop the universe from expanding and the universe will collapse back on itself—assuming that there is no repulsive force such as a cosmological constant. Conversely, if the density of the universe is less than the critical density, the universe will continue to expand and the gravitational pull will not be enough to stop the universe from expanding. This scenario would result in the "Big Freeze", *where the universe cools as it expands and reaches a state of entropy. One theory proposes that the universe could collapse to the state where it began and then initiate another Big Bang so in this way the universe would last forever, but would pass through phases of expansion (Big Bang) and contraction (Big Crunch)*

Recent experimental evidence (namely the observation of distant supernovae as standard candles, and the well-resolved mapping of the cosmic microwave background) has led to speculation that the expansion of the universe is not being slowed down by gravity but rather accelerating.

However, since the nature of the dark energy that is postulated to drive the acceleration is unknown, it is still possible (though not observationally supported as of today) that it might eventually reverse its developmental path and cause a collapse.

With the hypothesis that dark energy (De) is derived from dark matter (Dm) and is quantified by the Holy Grail equation, it can now be said that the observable universe (OU) will indeed cease to expand when the Dm is depleted. Gravity will then take over and the universal expansion will reverse itself.

Currently there are ~ 10^{183} STEPs in the OU, 10^{80} from the original STENs and 10^{60} shroud Steps plus 10^{43} from the expansion of the OU since the big bang. When the OU reaches 10^{200} STEPS, it will have expanded 40% of the quantum theory allowable restriction of 10^{500} possible units of interest.

Therefore, when the OU doze reach the maximum 10^{500} STEPs, the Dm will have been exhausted and the big crunch will begin by gravity gaining the upper hand.

Part V

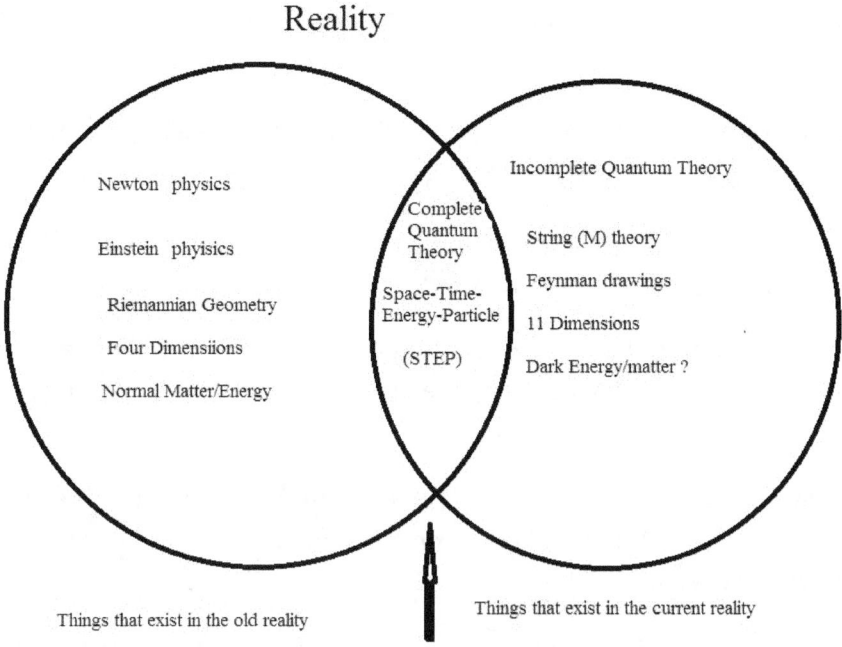

Reality

Newton physics

Einstein phyisics

Riemannian Geometry

Four Dimensiions

Normal Matter/Energy

Complete Quantum Theory

Space-Time-Energy-Particle

(STEP)

Incomplete Quantum Theory

String (M) theory

Feynman drawings

11 Dimensions

Dark Energy/matter ?

Things that exist in the old reality

Things that exist in the current reality

Things that exist in the new reality

The New Reality

"Reality is merely an illusion, albeit a very persistent one."

The new reality is much like the old reality before quantum theory set in.

This new reality is a paradigm shift.

The basic difference between the new reality and quantum reality is that everything that quantum theory predicts is tempered by the mechanism of STEP synchronization.

Take for example Brian Green's microscopic weirdness in the H-bar where he discusses Grace and George walking through a painted door on a wall in the tavern. He states that this would be a normal event at the microscopic level of our universe. This, he says is because quantum theory predicts it.

He relies on the fact that quantum theory has never failed in any of its predictions.

To date I've never heard of any one walking through a wall in my life time or *any other time for that matter.

But Green says at the microscopic (or Planck's scale) things are different because quantum theory says they are.

Let me clearly state that I think quantum theory (if rendered complete) will always predict with great accuracy the probability of things happening………. "whatever can happen --will happen"

The driver here is the word "can". If you do not take into account the effect that synchronized STEPs would have on a system, then the weirdness in the H-bar would prevail.

However, when STEP synchronization is considered, a different reality results, ----THE NEW REALITY.

Grace and George could only pass through the painted door only if the STEPs associate with the painted door alien by synchronization to allow it to happen.

Einstein was right, quantum theory is incomplete since it predicts without the effect STEPs have on its results.

Einstein and even Niels Bohr were skeptics of quantum results.

Einstein called Quantum theory incomplete but during his lifetime he was not able to show a pathway to temper the quantum theory to fit into the reality that we live in.

Niels Borg said that "to understand quantum fully is not without becoming dizzy".

Stephen Hawking said "Einstein was confused, not the quantum theory".

I think not.

Let's take a look at the double slit experiment and Feynman's "sum-over-paths" approach to quantum mechanics.

The double slit experiment shows that an electron can act as a particle and also as a wave.

Earlier I explained that the Planck size string-like electron is only one Planck's length in size ($\sim 10^{-35}$ m) and would fit comfortably in a STEP touching both the entry port and the exit port, sometimes referred to as the black hole entry and the white hole exit. The volume of one cubic Planck's unit is $10^{-105} m^3$. However, when the electron is considered a point particle it has a diameter length of $10^{-15}m$ and is found in a volume of $\sim 10^{60}$ STEPs.

The electron string vibrates throughout this volume at $\sim 10^{43}$ tics per second. It moves about in this volume as a "standing wave". This volume is considered the shroud of an elementary particle. I also noted earlier that this electron shroud moves through all STEPs as a "traveling wave".

Note: Remember there are more Planck ticks (10^{43}) in a second than there are seconds since the big bang (10^{17} seconds)!

This alone should make you stop and think that an elementary string particle can be in many places in just one second ($\sim 10^{60}$ places). However, this concept can be expanded even further.

Remember the string is vibrating with kinetic energy and thus the shroud has an effect on the neighboring STEPs. This in effect would allow the standing wave concept of the shroud along with the traveling wave associated with travel movement to account for the appearance that the electron was in two places at the same time. The single electron could move through one slit but the waves associated with it would pass through both. The lines of wave cancelation would indicate that the electron entered through both slits simultaneously, but in actuality only the standing waves of the shroud passed through one slit and the traveling wave disturbance entered the other slit. It was these two waves associated with the electron that actually canceled each other out and not the electron being in two places at one time.

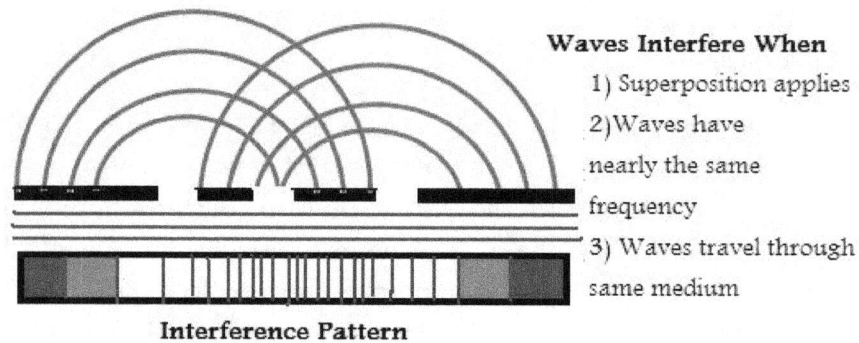

Waves Interfere When
1) Superposition applies
2) Waves have nearly the same frequency
3) Waves travel through same medium

Interference Pattern

Fig#24. The electron acts similar to a photon because both have the same wave action. The major difference between them is that an electron has measurable mass and a photon doesn't. An electron has a standing wave as well as travel wave action whereas the photon is not only a travel wave but also is a quantum particle as proven by Einstein's photo electric effect discovery.

Also the concept that an object is only in a location when looked at needs to be addressed by the complete quantum theory. See below **Wikipedia's** discussion on this.

Heisenberg's' uncertainty principal is often confused with the observer effect. See the Wikipedia discussion below;

*Heisenberg's **uncertainty principle**, is any of a variety of mathematical inequalities asserting a fundamental limit to the precision with which certain pairs of physical properties of a particle known as complementary variables.*

The uncertainty principle has been frequently confused with the observer effect, evidently even by its originator, Werner Heisenberg. The uncertainty principle in its standard form actually describes how precisely we may measure the position and momentum of a particle at the same time — if we increase the precision in measuring one quantity, we are forced to lose precision in measuring the other. An alternative version of the uncertainty principle, more in the spirit of an observer effect, fully accounts for the disturbance the observer has on a system and the error incurred, although this is not how the term "uncertainty principle" is most commonly used in practice.

Historically, the uncertainty principle has been confused with a somewhat similar effect in physics, called the observer effect, which notes that measurements of certain systems cannot be made without affecting the systems. Heisenberg offered such an observer effect at the quantum level (see below) as a physical "explanation" of quantum uncertainty. It has since become clear, however, that the uncertainty principle is inherent in the properties of all wave-like systems and that it arises in quantum mechanics simply due to the matter wave nature of all quantum objects. Thus, the uncertainty principle actually states a fundamental property of quantum systems, and is not a statement about the observational success of current technology.

This observer effect has lead people to believe that an object is only there when looked at.

When an electron is summoned to go from point A to point B, such as from an electron gum to a collector screen, the electron has more than one path it can take without violating any distance per time requirement.

When a double slit barrier is between point A and point B the STEPs will align themselves for the electron to pass through in a number of ways, much like the "sum over path ways" that Feynman proposed.

However sometimes an alternate pathway can be obstructed as in the case of an observer (i.e. a detector on a double slit set up). When the detector is turned off the electron can take path A or B. But when the detector is turned on the alternate pathway is consumed by the detector's photons thus leaving only the primarily pathway to be utilized by the electron in rout through the slit.

This is an example of how the observer effect would work in the double slit

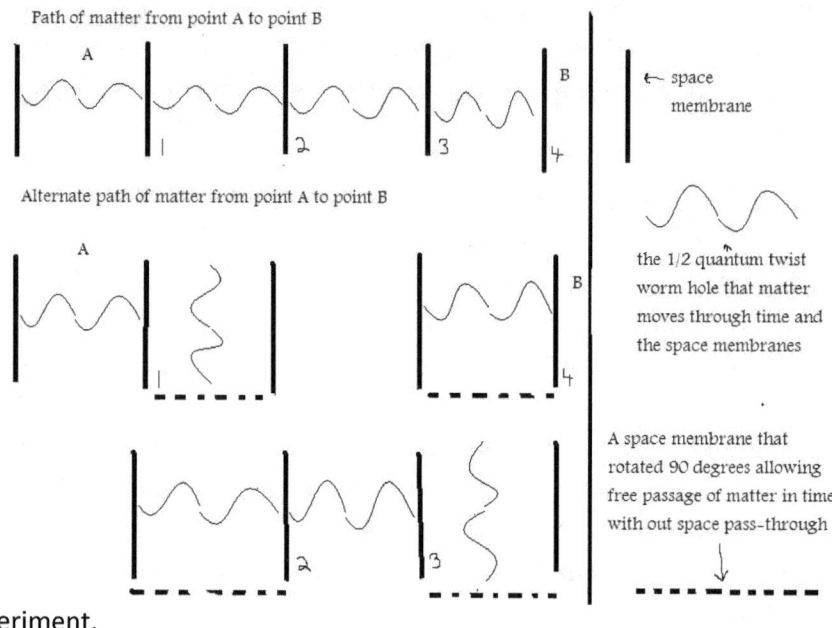

experiment.

Fig#25. The synchronization of STEPs allow matter to travel by many routes and still move at the same rate. Matter travels through 4 space membranes from point A to point B by either path.

Grace and George walking through a painted door on a wall in the tavern suffer the same consequence.

Concentrated matter has a much less chance for STEPs realignment than a single electron or photon.

However, in all cases, single or a multi-matter, the requirement of STEP synchronization is mandatory for travel or lack of it to take place.

The number of possible STEPs from the ~10^{183} found in our observable universe allows for a near infinite possibility of paths it can travel. Feynman uses the "sum-over-paths" to calculate the rout (es) of the electron which is well in agreement with the wave function results.

This weirdness of electron travel by quantum calculations and "sum-over-paths" can also be simply explained by STEP synchronization.

The synchronization of STEPs allow matter to get to its target without effecting the time it takes to get there.

Remember normal matter (bosons or fermions) only travels through space at the speed of light or less, but they can move through time at superluminal/instantaneous speed. (See the earlier image Fig# 09 and Fig#25).

Distance traveled is only the passage of matter through time and the space membrane.

As seen here when matter travels from point A to point B, it passes through 4 space membranes whether it goes on a straight projector or on an alternate route.

This is because, when matter moves only in time but not through space, it is super luminous (instantaneous).

This explains a number of quantum weirdness principles, such as the "sum-over-paths" as discussed above since now this can be explained by the principle that the synchronized STEPs will automatically and instantaneously calculate the "best route" to take with possible rotated STEPs along the path way for such a journey.

This New Reality is the forging of STEPs with incomplete quantum theory to produce a complete framework to explain, reduce, eliminate, reevaluate, our current thinking.

The Complete Quantum Theory for a New Reality

In the New Reality H-bar cafe' the patrons would only be able to walk through a wall if quantum predicted it and STEPs allowed it, then and only then would it happen.

Another weird quantum prediction is about what Einstein referred to as "spooky action at a distance" To best layout this controversy I will use the development of the **EPR Paradox, as outlined by Wikipedia (as of 5/13/2014).**

"The EPR paradox is an early and influential critique leveled against the Copenhagen interpretation of quantum mechanics. Albert Einstein and his colleagues Boris Podolsky and Nathan Rosen (known collectively as EPR) designed a thought experiment which revealed that the accepted formulation of quantum mechanics had a consequence which had not previously been noticed, but which looked unreasonable at the time. The scenario described involved the phenomenon that is now known as quantum entanglement.

According to quantum mechanics, under some conditions, a pair of quantum systems may be described by a single wave function, which encodes the probabilities of the outcomes of experiments that may be performed on the two systems, whether jointly or individually. At the time the EPR article discussed below was written, it was known from experiments that the outcome of an experiment sometimes cannot be uniquely predicted. An example of such indeterminacy can be seen when a beam of light is incident on a half-silvered mirror. One half of the beam will reflect, the other will pass. If the intensity of the beam is reduced until only one photon is in transit at any time, whether that photon will reflect or transmit cannot be predicted quantum mechanically."

The short answer for this is that a photon will reflect if the synchronization of any of the $\sim 10^{183}$ STEPS in the observable universe takes that route, if it transmits it, it is based on the same logic. Because

132

STEPs are so small 10^{-35} m our current detection systems are unable to predict which path a photon may take-----thus Heisenberg's' uncertainty principle is inserted into the thinking for this "weird" phenomena.

The routine explanation of this effect was, at that time, provided by Heisenberg's' uncertainty principle. Physical quantities come in pairs which are called conjugate quantities. Examples of such conjugate pairs are position and momentum of a particle and components of spin measured around different axes. When one quantity was measured, and became determined, the conjugated quantity became indeterminate.

Heisenberg's' explained this as a disturbance caused by measurement.

The EPR paper, written in 1935, was intended to illustrate that this explanation is inadequate. It considered two entangled particles, referred to as A and B, and pointed out that measuring a quantity of a particle A will cause the conjugated quantity of particle B to become undetermined, even if there was no contact, no classical disturbance. The basic idea was that the quantum states of two particles in a system cannot always be decomposed from the joint state of the two."

"Heisenberg's' principle was an attempt to provide a classical explanation of a quantum effect sometimes called non-locality. According to EPR there were two possible explanations. Either there was some interaction between the particles, even though they were separated, or the information about the outcome of all possible measurements was already present in both particles.

The EPR authors preferred the second explanation according to which that information was encoded in some 'hidden parameters'. The first explanation, that an effect propagated instantly, across a distance, is in conflict with the theory of relativity. They then concluded that quantum mechanics was incomplete since, in its formalism, there was no room for such hidden parameters.

EPR and the Copenhagen interpretation both believed that entanglement takes place but Einstein basically opted out of the "there was some instantaneous interaction between the particles, even though they were separated" because matter and information cannot travel faster than the speed of light. EPR postulated that an inscribed DNA-like info might be in play.

However, since the synchronization of STEPs is instantaneous, its (synchronization) does not violate the speed of light because, like a friction-free clock, when one gear move all other gears associated with it move instantaneously.

This can be thought of as STEPs being the hidden variables which are not in the matter but in the space-time-energy particles (the DNA-like).

Since I have postulated that there are $\sim 10^{183}$ STEPs in the observable universe it can be concluded that information at the Planck level is instantaneous without violating Einstein's "speed of light rule."

Therefore, when STEPs are incorporated into the quantum theory it becomes complete which now satisfies Einstein's requirement for hidden variables and completeness.

Violations of the conclusions of Bell's theorem are generally understood to have demonstrated that the hypotheses of Bell's theorem, also assumed by Einstein, Poldolsky and Rosen, do not apply in our world. Most physicists who have examined the issue concur that experiments, such as those of Alain Aspect and his group, have confirmed that physical probabilities, as predicted by quantum theory, do exhibit the phenomena of Bell- inequality violations that are considered to invalidate EPR's preferred "local hidden-variables" type of explanation for the correlations to which EPR first drew attention."

Again when STEPs are incorporated the Bell's theorem is void, because the hidden variable is in space-time and not in the matter itself.

The article that first brought forth these matters, "Can Quantum-Mechanical Description of Physical Reality Be Considered Complete?" was published in 1935. Einstein struggled to the end of his life for a theory that could better comply with his idea of causality, protesting against the view that there exists no objective physical reality other than that which is revealed through measurement interpreted in terms of quantum mechanical formalism. However, since Einstein's death, experiments analogous to the one described in the EPR paper have been carried out, starting in 1976 by French scientists Lamehi-Rachti and Mittig at the Saclay Nuclear Research Centre. These experiments appear to show that the local realism idea is false.

However, with the advent of STEP synchronization incorporated into Bell's theorem, the local realism can no longer be considered false.

Quantum mechanics and its interpretation Main article: Interpretations of quantum mechanics since the early twentieth century, quantum theory has proved to be successful in describing accurately the physical reality of the mesoscopic and microscopic world, in multiple reproducible physics experiments.

Quantum mechanics was developed with the aim of describing atoms and explaining the observed spectral lines in a measurement apparatus. Although disputed, it has yet to be seriously challenged. Philosophical interpretations of quantum phenomena, however, are another matter: the question of how to interpret the mathematical formulation of quantum mechanics has given rise to a variety of different answers from people of different philosophical persuasions (see Interpretations of quantum mechanics).

Quantum theory and quantum mechanics do not provide single measurement outcomes in a deterministic way.

According to the understanding of quantum mechanics known as the Copenhagen interpretation, measurement causes an instantaneous collapse of the wave function describing the quantum system into an eigenstate of the observable that was measured. Einstein characterized this imagined collapse in the 1927 Solvay Conference. He presented a thought experiment in which electrons are introduced through a small hole in a sphere whose inner surface serves as a detection screen.

The electrons will contact the spherical detection screen in a widely dispersed manner. Those electrons, however, are all individually described by wave fronts that expand in all from the point of entry. A wave as it is understood in everyday life would paint a large area of the detection screen, but the electrons would be found to impact the screen at single points and would eventually form a pattern in keeping with the probabilities described by their identical wave functions. Einstein asks what makes each electron's wave front "collapse" at its respective location. Why do the electrons appear as single bright scintillations rather than as dim washes of energy across the surface? Why does any single electron appear at one point rather than some alternative point? The behavior of the electrons gives the impression of some signal having been sent to all possible points of contact that would have nullified all but one of them, or, in other words, would have preferentially selected a single point to the exclusion of all others.

Einstein's opposition

Einstein was the most prominent opponent of the Copenhagen interpretation. In his view, quantum mechanics is incomplete. Commenting on this, other writers (such as John von Neumann and David Bohm hypothesized) that consequently there would have to be 'hidden' variables responsible for random measurement results, something which was not expressly claimed in the original paper.

STEPs would have helped here if only Einstein was enlightened about them.

The 1935 EPR paper condensed the philosophical discussion into a physical argument. The authors claim that given a specific experiment, in which the outcome of a measurement is known before the measurement takes place, there must exist something in the real world, an "element of reality", that determines the measurement outcome. They postulate that these elements of reality are local, in the sense that each belongs to a certain point in space-time. Each element may only be influenced by events which are located in the backward light cone of its point in space-time (i.e., the past).

136

These claims are founded on assumptions about nature that constitute what is now known as local realism.

Though the EPR paper has often been taken as an exact expression of Einstein's views, it was primarily authored by Podolsky, based on discussions at the Institute for Advanced Study with Einstein and Rosen. Einstein later expressed to Erwin Schrödinger that, "it did not come out as well as I had originally wanted; rather, the essential thing was, so to speak, smothered by the formalism." In 1936, Einstein presented an individual account of his local realist ideas.

Description of the paradox

The original EPR paradox challenges the prediction of quantum mechanics that it is impossible to know both the position and the momentum of a quantum particle. This challenge can be extended to other pairs of physical properties.

EPR paper

The original paper purports to describe what must happen to "two systems I and II, which we permit to interact.", and, after some time, "we suppose that there is no longer any interaction between the two parts." In the words of Kumar (2009), the EPR description involves "two particles, A and B, [which] interact briefly and then move off in opposite directions." According to Heisenberg's' uncertainty principle, it is impossible to measure both the momentum and the position of particle B exactly. However, according to Kumar, it is possible to measure the exact position of particle A. By calculation, therefore, with the exact position of particle A known, the exact position of particle B can be known. Also, the exact momentum of particle A can be measured, so the exact momentum of particle B can be worked out. Kumar writes: "EPR argued that they had proved that ... [particle] B can have simultaneously exact values of position and momentum. ...

Particle B has a position that is real and a momentum that is real."

EPR appeared to have contrived a means to establish the exact values of either the momentum or the position of B due to measurements made on *particle A, without the slightest possibility of particle B being physically disturbed.*

EPR tried to set up a paradox to question the range of true application of Quantum Mechanics: Quantum theory predicts that both values cannot be known for a particle, and yet the EPR thought experiment purports to show that they must all have determinate values. The EPR paper says: "We are thus forced to conclude that the quantum-mechanical description of physical reality given by wave functions is not complete."

The EPR paper ends by saying: While we have thus shown that the wave function does not provide a complete description of the physical reality, we left open the question of whether or not such a description exists. We believe, however, that such a theory is possible.

The STEP theory coupled with quantum theory allow for a complete description.

Measurements on an entangled state

We have a source that emits electron–positron pairs, with the electron sent to destination A, where there is an observer named Alice, and the positron sent to destination B, where there is an observer named Bob. According to quantum mechanics, we can arrange our source so that each emitted pair occupies a quantum state called a spin singlet. The particles are thus said to be entangled. This can be viewed as a quantum superposition of two states, which we call state I and state II. In state I, the electron has spin pointing upward along the z-axis (+z) and the positron has spin pointing downward along the z-axis (–z). In state II, the electron has spin –z and the positron has spin +z. Therefore, it is impossible (without measuring) to know the definite state of spin of either particle in the spin singlet.

Alice now measures the spin along the z-axis. She can obtain one of two possible outcomes: +z or –z. Suppose she gets +z. According to the Copenhagen interpretation of quantum mechanics, the quantum state of the system collapses into state I. The quantum state determines the probable outcomes of any measurement performed on the system. In this case, if Bob subsequently measures spin along the z-axis, there is 100% probability that he will obtain −z. Similarly, if Alice gets −z, Bob will get +z.

There is, of course, nothing special about choosing the z-axis: according to quantum mechanics the spin singlet state may equally well be expressed as a superposition of spin states pointing in the x direction. Suppose that Alice an

d Bob had decided to measure spin along the x-axis. We'll call these states Ia and IIa. In state Ia, Alice's electron has spin +x and Bob's positron has spin −x. In state IIa, Alice's electron has spin −x and Bob's positron has spin +x. Therefore, if Alice measures +x, the system 'collapses' into state Ia, and Bob will get −x. If Alice measures −x, the system collapses into state IIa, and Bob will get +x.

Whatever axis their spins are measured along, they are always found to be opposite. This can only be explained if the particles are linked in some way. Either they were created with a definite (opposite) spin about every axis—a "hidden variable" argument—or they are linked so that one electron "feels" which axis the other is having its spin measured along, and becomes its opposite about that one axis—an "entanglement" argument. Moreover, if the two particles have their spins measured about different axes, once the electron's spin has been measured about the x-axis (and the positron's spin about the x-axis deduced), the positron's spin about the z-axis will no longer be certain, as if (a) it knows that the measurement has taken place, or (b) it has a definite spin already, about a second axis—a hidden variable. However, it turns out that the predictions of Quantum Mechanics, which have been confirmed by experiment, cannot be explained by any hidden variable theory. This is demonstrated in Bell's theorem.

In quantum mechanics, the x-spin and z-spin are "incompatible observables", meaning the Heisenberg uncertainty principle applies to alternating measurements of them: a quantum state cannot possess a definite value for both of these variables.

Suppose Alice measures the z-spin and obtains +z, so that the quantum state collapses into state I. Now, instead of measuring the z-spin as well,

Bob measures the x-spin. According to quantum mechanics, when the system is in state I, Bob's x-spin measurement will have a 50% probability of producing +x and a 50% probability of -x. It is impossible to predict which outcome will appear until Bob actually performs the measurement.

Here is the crux of the matter. You might imagine that, when Bob measures the x-spin of his positron, he would get an answer with absolute certainty, since prior to this he hasn't disturbed his particle at all. Bob's positron has a 50% probability of producing +x and a 50% probability of –x—so the outcome is not certain. Bob's positron "knows" that Alice's electron has been measured, and its z-spin detected, and hence B's z-spin has been calculated, but the x-spin of Bob's positron remains uncertain.

Put another way, how does Bob's positron know which way to point if Alice decides (based on information unavailable to Bob) to measure x (i.e., to be the opposite of Alice's electron's spin about the x-axis) and also how to point if Alice measures z, since it is only supposed to know one thing at a time? The Copenhagen interpretation rules that say the wave function "collapses" at the time of measurement, so there must be action at a distance (entanglement) or the positron must know more than it's supposed to know (hidden variables).

The answer based on STEPs and Quantum theory is that synchronization of STEPs is the cause of entanglement and STEPs themselves are the hidden variables.

Here is the paradox summed up:

It is one thing to say that physical measurement of the first particle's momentum affects uncertainty in its own position, but to say that measuring the first particle's momentum affects the uncertainty in the position of the other is another thing altogether. Einstein, Podolsky and Rosen asked how can the second particle "know" to have precisely defined momentum but uncertain position? Since this implies that one particle is communicating with the other instantaneously across space, i.e., faster than light, this is the "paradox".

Incidentally, Bell used spin as his example, but many types of physical quantities—referred to as "observables" in quantum mechanics—can be used. The EPR paper used momentum for the observable. Experimental realizations of the EPR scenario often use photon polarization, because polarized photons are easy to prepare and measure.

Locality in the EPR experiment

The principle of locality states that physical processes occurring at one place should have no immediate effect on the elements of reality at another location. At first sight, this appears to be a reasonable assumption to make, as it seems to be a consequence of special relativity, which states that information can never be transmitted faster than the speed of light without violating causality. It is generally believed that any theory which violates causality would also be internally inconsistent, and thus useless.

It turns out that the usual rules for combining quantum mechanical and classical descriptions violate the principle of locality without violating causality. Causality is preserved because there is no way for Alice to transmit messages (i.e., information) to Bob by manipulating her measurement axis. Whichever axis she uses, she has a 50% probability of obtaining "+" and 50% probability of obtaining, completely at random; according to quantum mechanics, it is fundamentally impossible for her to influence what result she gets. Furthermore, Bob is only able to perform his measurement once: there is a fundamental property of quantum mechanics, known as the "no cloning theorem", which makes it impossible for him to make a million copies of the electron he receives, perform a spin measurement on each, and look at the statistical distribution of the results. Therefore, in the one measurement he is allowed to make, there is a 50% probability of getting "+" and 50% of getting "−", regardless of whether or not his axis is aligned with Alice's.

However, the principle of locality appeals powerfully to physical intuition, and Einstein, Podolsky and Rosen were unwilling to abandon it. Einstein derided the quantum mechanical predictions as "spooky action at a distance". The conclusion they drew was that quantum mechanics is not a complete theory.

141

Their conclusion is valid with the knowledge that STEPs make the quantum theory complete.

In recent years, however, doubt has been cast on EPR's conclusion due to developments in understanding locality and especially quantum decoherence. The word locality has several different meanings in physics. For example, in quantum field theory "locality" means that quantum fields at different points of space do not interact with one another. However, quantum field theories that are "local" in this sense appear to violate the principle of locality as defined by EPR, but they nevertheless do not violate locality in a more general sense. Wavefunction collapse can be viewed as an epiphenomenon of quantum decoherence, which in turn is nothing more than an effect of the underlying local time evolution of the wavefunction of a system and all of its environment. Since the underlying behaviour doesn't violate local causality, it follows that neither does the additional effect of wavefunction collapse, whether real or apparent. Therefore, as outlined in the example above, neither the EPR experiment nor any quantum experiment demonstrates that faster-than-light signaling is possible.

Resolving the paradox Hidden variables

There are several ways to resolve the EPR paradox. The one suggested by EPR is that quantum mechanics, despite its success in a wide variety of experimental scenarios, is actually an incomplete theory. In other words, there is some yet undiscovered theory of nature to which quantum mechanics acts as a kind of statistical approximation (albeit an exceedingly successful one). Unlike quantum mechanics, the more complete theory contains variables corresponding to all the "elements of reality". There must be some unknown mechanism acting on these variables to give rise to the observed effects of "non-commuting quantum observables", i.e. the Heisenberg uncertainty principle. Such a theory is called a hidden variable theory.

Need I explain? Or should we rely on uncertainties.

To illustrate this idea, we can formulate a very simple hidden variable theory for the above thought experiment. One supposes that the quantum spin-singlet states emitted by the source are actually approximate descriptions for "true" physical states possessing definite values for the z-spin and x-spin. In these "true" states, the positron going to Bob always has spin values opposite to the electron going to Alice, but the values are otherwise completely random. For example, the first pair emitted by the source might be "(+z, −x) to Alice and (−z, +x) to Bob", the next pair "(−z, −x) to Alice and (+z, +x) to Bob", and so forth. Therefore, if Bob's measurement axis is aligned with Alice's, he will necessarily get the opposite of whatever Alice gets; otherwise, he will get "+" and "−" with equal probability.

Assuming we restrict our measurements to the z- and x axes, such a hidden variable theory is experimentally indistinguishable from quantum mechanics. In reality, there may be an infinite number of axes along which Alice and Bob can perform their measurements, so there would have to be an infinite number of independent hidden variables. However, this is not a serious problem; we have formulated a very simplistic hidden variable theory, and a more sophisticated theory might be able to patch it up. It turns out that there is a much more serious challenge to the idea of hidden variables.

Seriously? Only when STEPs are not considered.

Bell's inequality

Main article: Bell's theorem

In 1964, John Bell showed that the predictions of quantum mechanics in the EPR thought experiment are significantly different from the predictions of a particular class of hidden variable theories (the local hidden variable theories).

Roughly speaking, quantum mechanics has a much stronger statistical correlation with measurement results performed on different axes than do these hidden variable theories. These differences, expressed using inequality relations known as "Bell's inequalities", are in principle experimentally detectable. Later work by Eberhard showed that the key properties of local hidden variable theories which lead to Bell's inequalities are locality and counter-factual definiteness. Any theory in which these principles apply produces the inequalities. Arthur Fine subsequently showed that any theory satisfying the inequalities can be modeled by a local hidden variable theory.

After the publication of Bell's paper, a variety of experiments to test Bell's inequalities were devised. These generally relied on measurement of photon polarization. All experiments conducted to date have found behavior in line with the predictions of standard quantum mechanics theory.

However, Bell's theorem does not apply to all possible philosophically realist theories. It is a common misconception that quantum mechanics is inconsistent with all notions of philosophical realism. Realist interpretations of quantum mechanics are possible, although as discussed above, such interpretations must reject either locality or counter-factual definiteness. Mainstream physics prefers to keep locality, while striving also to maintain a notion of realism that nevertheless rejects counter-factual definiteness. Examples of such mainstream realist interpretations are the consistent histories interpretation and the transactional interpretation (first proposed by John G. Cramer in 1986). Fine's work showed that, taking locality as a given, there exist scenarios in which two statistical variables are correlated in a manner inconsistent with counter-factual definiteness, and that such scenarios are no more mysterious than any other, despite the fact that the inconsistency with counter-factual definiteness may seem 'counter-intuitive'. Violation of locality is difficult to reconcile with special relativity, and is thought to be incompatible with the principle of causality. However, the Bohm interpretation of quantum mechanics keeps counter-factual definiteness, while introducing a conjectured non-local mechanism in the form of the 'quantum potential' that is defined as one of the terms of the Schrödinger equation. Some workers in the field have also attempted to formulate hidden variable theories that exploit loopholes in actual experiments, such as the assumptions made in interpreting experimental data, although no theory has been proposed that can reproduce all the results of quantum mechanics.

There are also individual EPR-like experiments that have no local hidden variables explanation. Examples have been suggested by David Bohm and by Lucien Hardy.

All this (Bell's Theorems) becomes a moot point with the knowledge of STEPs incorporated into the quantum theory.

Einstein's hope for a purely algebraic theory

The Bohm interpretation of quantum mechanics hypothesizes that the state of the universe evolves smoothly through time with no collapsing of quantum wavefunctions. One problem for the Copenhagen interpretation is to precisely define wavefunction collapse. Einstein maintained that quantum mechanics is physically incomplete and logically unsatisfactory. In "The Meaning of Relativity", Einstein wrote, "One can give good reasons why reality cannot at all be represented by a continuous field. From the quantum phenomena it appears to follow with certainty that a finite system of finite energy can be completely described by a finite set of numbers (quantum numbers). This does not seem to be in accordance with a continuum theory and must lead to an attempt to find a purely algebraic theory for the representation of reality. But nobody knows how to find the basis for such a theory." If time, space, and energy are secondary features derived from a substrate below the Planck scale, then Einstein's hypothetical algebraic system might resolve the EPR paradox (although Bell's theorem would still be valid). Edward Fredkin in the Fredkin Finite Nature Hypothesis has suggested an informational basis for Einstein's hypothetical algebraic system. If physical reality is totally finite, then the Copenhagen interpretation might be an approximation to an information processing system below the Planck scale.

"Acceptable theories" and the experiment according to the present view of the situation, quantum mechanics flatly contradicts Einstein's philosophical postulate that any acceptable physical theory must fulfill "local realism".

In the EPR paper (1935), the authors realized that quantum mechanics was inconsistent with their assumptions, but Einstein nevertheless thought that quantum mechanics might simply be augmented by hidden variables (i.e., variables which were, at that point, still obscure to him), without any other change, to achieve an acceptable theory. He pursued these ideas for over twenty years until the end of his life, in 1955.

These revelations of space, time, energy are the answer to Einstein's lifelong pursue for an algebraic theory that hovers around the Planck's size synchronization of the universe and on to the complete observable universe we live in.

In contrast, John Bell, in his 1964 paper, showed that quantum mechanics and the class of hidden variable theories Einstein favored would lead to different experimental results: different by a factor of 3/2 for certain correlations. So the issue of "acceptability", up to that time mainly concerning theory, finally became experimentally decidable.

There are many Bell test experiments, e.g., those of Alain Aspect and others. They support the predictions of quantum mechanics rather than the class of hidden variable theories supported by Einstein. According to Karl Popper, these experiments showed that the class of "hidden variables" Einstein believed in is erroneous.

Implications for quantum mechanics

Most physicists today believe that quantum mechanics is correct, and that the EPR paradox is a "paradox" only because classical intuitions do not correspond to physical reality.

They need to apply STEP synchronization to their thought experiments.

How EPR is interpreted regarding locality depends on the interpretation of quantum mechanics one uses. In the Copenhagen interpretation, it is usually understood that instantaneous wave function collapse does occur. However, the view that there is no causal instantaneous effect has also been proposed within the Copenhagen interpretation: in this alternate view, measurement affects our ability to define (and measure) quantities in the physical system, not the system itself. In the many-worlds interpretation, locality is strictly preserved, since the effects of operations such as measurement affect only the state of the particle that is measured.

146

However, the results of the measurement are not unique—every possible result is obtained.

The EPR paradox has deepened our understanding of quantum mechanics by exposing the fundamentally non classical characteristics of the measurement process. Before the publication of the EPR paper, a measurement was often visualized as a physical disturbance inflicted directly upon the measured system. For instance, when measuring the position of an electron, one imagines shining a light on it, thus disturbing the electron and producing the quantum mechanical uncertainties in its position. Such explanations, which are still encountered in popular expositions of quantum mechanics, are debunked by the EPR paradox, which shows that a "measurement" can be performed on a particle without disturbing it directly, by performing a measurement on a distant entangled particle. In fact, Yakir Aharonov and his collaborators have developed a whole theory of so-called Weak measurement.

Technologies relying on quantum entanglement are now being developed. In quantum cryptography, entangled particles are used to transmit signals that cannot be eavesdropped upon without leaving a trace. In quantum computation, entangled quantum states are used to perform computations in parallel, which may allow certain calculations to be performed much more quickly than they ever could be with classical computers.

Hopefully, the complete theory of quantum with the space, time and energy entanglement will enhance communication.

The unparticle-theoretical view

Not all theoretical Physicist are believers of the particle physics theory.

Dr. Howard Georgi of Harvard University has scraped the particle theory for a factorial point of view about how the universe is held together. He envisions a tree as a factorial component of space- time, and that the leaves on the tree are matter that is held in place by the tree branches and trunk.

This theory is not too far-off of an idea from my STEPs theory since both the factorial tree branches and the STEPs have similar functions when associated with matter. The tree branches hold the matter (leaves) in place while the STEPs are synchronized to allow matter to flow, but also like the tree the STEPs are entangled with matter.

The specialized Steps, the STENs (10^{80} that were created prior to inflation) act as huge scaffold that hold the galaxies in place.

Where-as Dr. Georgi abandon particles for the factorial arrangement of space-time, I embrace them by the incorporation of ~10^{183} Space-Time-Energy-Particles in our observable universe.

Possibly if Dr. Georgi could incorporate STEPs in his equations, his theory could be advanced and the synchronization of the STEPs can be shown.

Dark matter and dark energy is discussed by NASA as shown in the following italicized print.

http://science.nasa.gov/astrophysics/focus-areas/what-is-dark

Dark Matter

By fitting a theoretical model of the composition of the Universe to the combined set of cosmological observations, scientists have come up with the composition that we described above, ~68% dark energy, ~27% dark matter, ~5% normal matter. What is dark matter?

We are much more certain what dark matter is not than we are what it is. First, it is dark, meaning that it is not in the form of stars and planets that we see. Observations show that there is far too little visible matter in the Universe to make up the 27% required by the observations. Second, it is not in the form of dark clouds of normal matter, matter made up of particles called baryons. We know this because we would be able to detect baryonic clouds by their absorption of radiation passing through them. Third, dark matter is not antimatter, because we do not see the unique gamma rays that are produced when antimatter annihilates with matter. Finally, we can rule out large galaxy-sized black holes on the basis of how many gravitational lenses we see. High concentrations of matter bend light passing near them from objects further away, but we do not see enough lensing events to suggest that such objects to make up the required 25% dark matter contribution. However, at this point, there are still a few dark matter possibilities that are viable. Baryonic matter could still make up the dark matter if it were all tied up in brown dwarfs or in small, dense chunks of heavy elements. These possibilities are known as massive compact halo objects, or "MACHOs ". But the most common view is that dark matter is not baryonic at all, but that it is made up of other, more exotic particles like axions or WIMPS (Weakly Interacting Massive Particles).

I do not believe that dark matter is made from baryonic matter or WIMPS! As discussed throughout this book, I believe that dark matter is found in the 10^{80} STENs scattered throughout the universe, mainly near normal matter like galaxies etc.

This dark matter is the reservoir from which dark energy is derived. Most likely the conversion is similar to normal energy and normal matter exchange by $E=MC^2$.

I propose $De=DmC^2$. De and Dm are dark energy and dark matter respectively.

The basic concept of the exchange of dark matter to dark energy by a related speed factor of C^2 is what I hypothesize.

Dark Energy

In the early 1990's, one thing was fairly certain about the expansion of the Universe. It might have enough energy density to stop its expansion and recollapse, it might have so little energy density that it would never stop expanding, but gravity was certain to slow the expansion as time went on. Granted, the slowing had not been observed, but, theoretically, the Universe had to slow. The Universe is full of matter and the attractive force of gravity pulls all matter together. Then came 1998 and the Hubble Space Telescope (HST) observations of very distant supernovae that showed that, a long time ago, the Universe was actually expanding more slowly than it is today. So the expansion of the Universe has not been slowing due to gravity, as everyone thought, it has been accelerating. No one expected this, no one knew how to explain it. But something was causing it.

Eventually theorists came up with three sorts of explanations. Maybe it was a result of a long-discarded version of what was called a "cosmological constant." Maybe there was some strange kind of energy-fluid that filled space. Maybe there is something wrong with Einstein's theory of gravity and a new theory could include some kind of field that creates this cosmic acceleration. Theorists still don't know what the correct explanation is, but they have given the solution a name. It is called dark energy.

What is Dark Energy?

More is unknown than is known. We know how much dark energy there is because we know how it affects the Universe's expansion. Other than that, it is a complete mystery. But it is an important mystery. It turns out that roughly 68% of the Universe is dark energy. Dark matter makes up about 27%. The rest - everything on Earth, everything ever observed with all of our instruments, all normal matter - ads up to less than 5% of the Universe. Come to think of it, maybe it shouldn't be called "normal" matter at all, since it is such a small fraction of the Universe.

One explanation for dark energy is that it is a property of space. Albert Einstein was the first person to realize that empty space is not nothing. Space has amazing properties, many of which are just beginning to be understood. The first property that Einstein discovered is that it is possible for more space to come into existence. Then one version of Einstein's gravity theory, the version that contains a cosmological constant, makes a second prediction: "empty space" can possess its own energy.

Because this energy is a property of space itself, it would not be diluted as space expands. As more space comes into existence, more of this energy-of-space would appear. As a result, this form of energy would cause the Universe to expand faster and faster. Unfortunately, no one understands why the cosmological constant should even be there, much less why it would have exactly the right value to cause the observed acceleration of the Universe.

Another explanation for how space acquires energy comes from the quantum theory of matter. In this theory, "empty space" is actually full of temporary ("virtual") particles that continually form and then disappear. But when physicists tried to calculate how much energy this would give empty space, the answer came out wrong - wrong by a lot. The number came out 10^{120} times too big. That's a 1 with 120 zeros after it. It's hard to get an answer that bad. So the mystery continues.

Another explanation for dark energy is that it is a new kind of dynamical energy fluid or field, something that fills all of space but something whose effect on the expansion of the Universe is the opposite of that of matter and normal energy. Some theorists have named this "quintessence," after the fifth element of the Greek philosophers. But, if quintessence is the answer, we still don't know what it is like, what it interacts with, or why it exists. So the mystery continues.

A last possibility is that Einstein's theory of gravity is not correct. That would not only affect the expansion of the Universe, but it would also affect the way that normal matter in galaxies and clusters of galaxies behaved. This fact would provide a way to decide if the solution to the dark energy problem is a new gravity theory or not: we could observe how galaxies come together in clusters. But if it does turn out that a new theory of gravity is needed, what kind of theory would it be? How could it correctly describe the motion of the bodies in the Solar System, as Einstein's theory is known to do, and still give us the different prediction for the Universe that we need? There are candidate theories, but none are compelling. So the mystery continues.

The thing that is needed to decide between dark energy possibilities - a property of space, a new dynamic fluid, or a new theory of gravity - is more data, better data.

First and foremost, important- Einstein's theory of gravity is correct! Too many theoretical physicists want to prove Einstein wrong!

Go with the flow-advancing Einstein's fundamentals is a much more rewarding experience than trying to disprove him!

Secondly, if you want to think of dark energy to be some kind of a new dynamical fluid or field---think of STENs to be that fluid or field.

Thirdly, if you think of dark energy as being temperately virtual particles, then the math must be looked at differently.

The virtual particle pairs would actually be neighboring STEPs found next to string matter moving as traveling waves throughout space-time.

Next propose that 10^{60} STEPs surrounded the STENs which were produced during the inflationary period between ~10^{38} and ~10^{32} sec. contained the dark matter as virtual particles.

Remember, there are ~10^{183} STEP in the OU.

If you take the error of ~10^{120} from ~ 10^{183}(STEPs in OU), you get ~10^{63} (STEP), which is ~ the number of steps that surrounded STENs as shrouds when the universe was first expanded after the big bang.

These shroud STEPs would contain the virtual particles.

These ~$10^{\wedge}60$ virtual particle shroud STEPs would now give you a more accurate number than the ~$10^{\wedge}183$ OU would for estimation of dark energy.

However, in my hypothesis I theorize that the ~10^{120} STEPs contain the dark matter that converts to dark energy.

This is best understood when we use Einstein's gravity theory, the version that contains a cosmological constant.

Now if we subtract 10^{60} from 10^{183}, we get~ 10^{123} STEPs in the OU after we remove the shrouds containing string matter.

These 10^{120} STEPs would be a much closer accurate number than the whole ~10^{183} OU would for estimation of dark energy. See Part VI for further insight.

I propose that NADA check my theory out!

Remember when Einstein first came out with his relativity theories. There was not much fan-fair until an experiment was conducted to see if normal matter's gravity could bend light.

It was not until a total solar eclipse then Sir Arthur Eddington performed the first experimental test of Albert Einstein's general theory of relativity that drew attention to general relativity.

The results made Einstein a celebrity overnight, which precipitated in the eventual triumph of general relativity over classical Newtonian physics.

My theory is that dark matter is found as STENs and is converted to dark energy by a relationship of dark energy and dark matter similar to normal matter and normal energy, which is the cosmological constant that Einstein postulated.

An experiment to convert this thought experiment into an operational experiment that NADA could perform would be to compare the ratios of dark energy: dark matter at the early wmap, to the current wmap and a future wmap (after additional expansion) and confirm or expel the concept that dark matter is a concentrated form of dark energy and that by the conversion of dark matter into dark energy the OU expanse.

Many Worlds

From Wikipedia, the free encyclopedia

The many-worlds interpretation is an <u>interpretation of quantum mechanics</u> that asserts the <u>objective reality</u> of the <u>universal wavefunction</u> and denies the actuality of <u>wavefunction collapse</u>. Many-worlds implies that all <u>possible</u> alternate histories and futures are real, each representing an actual "world" (or "universe"). In lay terms, the hypothesis states there is a *very large—perhaps infinite—number of universes, and everything that could possibly have happened in our past, but did not, has occurred in the past of some other universe or universes.*

In many-worlds, the subjective appearance of wavefunction collapse is explained by the mechanism of quantum decoherence, and this is supposed to resolve all of the correlation paradoxes of quantum theory, such as the EPR paradox and Schrödinger's cat,[1] since every possible outcome of every event defines or exists in its own "history" or "world."

For the complete quantum theory to engulf the many worlds theory it would have to except the concept of decoherence. Whether it is or it isn't, is outside the realm of this discussion. What is in the realm is that if the many worlds exist they would be in the ~10^{183} present day STEP which should terminate at 10^{500} STEPs and as such would be considered outside of our upper limit of mathematical calculations as discussed earlier.

Part VI:

Holy Grail Equation of All Physics

$$Ef = Dm\Lambda_a$$

The search for the one single equation to describe all of the fundamentals of our observable universe (OU) is summed up in one small equation. This Equation is not too distant from some other small but powerful equations like F=ma and E=mC², where F is force, m is the mass of normal-matter (Nm) particle strings, and (a) represents acceleration. E is normal-energy (Ne) and C is the speed of light in a vacuum.

Actually these two basic equations relate to each other when you consider force to be energy and acceleration to be the speed of light squared; when F=E and a =c², then F=ma becomes E=mC².

For my Holy Grail Equation of Physics:

$$[Ef = Dm\Lambda_a]\ T,\ \Theta$$

Ef= Eternal force (some time referred to in literature as the super force), Dm = Dark matter (which I define as concentrated dark energy), Λ_a = ΛR/3 acceleration due to the cosmological constantΛ and relates to the vacuum energy density of the OU. R is the radius of the OU. Lambda Λ is based on the holographic method developed by G 'Hoot and L Susskind's black-hole entropy principle. T, Θ is the time and temperature of the OU when Ef is calculated.

Victor Stenger wrote in Reality Check, Skeptical Briefs Vol.21.1, 2011;

Physicist Leonard Susskind calls the problem with the cosmological constant "the mother of all physics problems" and "the worst prediction ever." The currently favored solution to the problem among physicists is called the "multiverse" in which our universe is just one of a great many *others having a wide variation of values for the cosmological constant as well as other physics parameters. We happen to live in the universe suitable for us. Susskind notes that string theory has some 10^{500} possible solutions, each of which could correspond to a separate universe within the multiverse.*

While I have nothing against the multiverse theory, my view is that the cosmological constant calculation is so obviously wrong that it can be ignored. While physicists have not yet reached a consensus on the correct calculation, one possibility that agrees with observations is called the holographic principle.

The calculation of the vacuum energy density of the universe involves a sum over all the zero-point energy states in the universe. The "worst prediction ever" assumed that the number of states is proportional to the volume, but now there is reason to believe that this is wrong. The holographic principle asserts that the number of states in a volume is proportional to the surface area of that volume, as in the case of a black hole. The universe can have no more states than that of a black hole of the same size. The energy density calculated from this assumption is of the same order of magnitude as the vacuum energy density that is determined from observations.

When the sum over volume is used for the quantum gravity for the cosmological constant, the *Holy Grail* Equation for all of Physics would be: **$[Ef=Dm\Lambda_a{}^{123}]$ T, Θ**, this density is 10^{123} times too much.

"Some scientists call the cosmological constant the "worst prediction of physics." When today's theories give an estimated value that is about 120 orders of magnitude larger than the measured value, it's hard to argue with that title." **http://phys.org/news/2012-03-weve-cosmological-constant-wrong.html#jCp**.

When the surface area of the OU is used instead of the volume, the exponent (123) drops out.

Some physicist postulate that only a God like creator could create a universe with such precision (10^{123}), others use the many worlds and random chance concepts to explain this.

My belief is; since all STEPs are all synchronized, the universe cannot be governed by random selection, it must be by design. This design includes the 10^{120} calibration factor for expansion as well as the ability for each STEP to respond to all others! Synchronization every 10^{43} Planck tics/second for 10^{183} STEPs with 10^{80} proton equivalent matter components, cannot be random, since each movement requires a nonrandom input. Also the quantum limit of 10^{500} events would be consumed in a very short time.

Note: the proof I use to justify that the STEPs are synchronized is based on the "spooky action at a distance" phenomena since the shared information of STEPs from one location in the OU to another in any corner or crevice is superluminal by way of "gears in a clock communication" as opposed to "pony express" communication.

Getting back to the subject at hand, I propose a third method to establish a reliable cosmological constant.

There are ~ 10^{183} total STEPs in the OU, but 10^{60} are consumed by normal-matter. Subtracting out - 10^{60} shroud STEPs where all normal-matter is confined ~ = 10^{123} remaining STEPs are left in a true vacuum.

The surface area of the entire OU is 10^{55} m^2 and according to Dr. Leonard Susskind, this surface area when used to calculate the density neatly matches the experimental density that satisfies the cosmological constant without a factor of 10^{123}.

When the surface area of the OU ($10^{\wedge}55$ m^2) is divided by the surface area of a single STEP (10^{-71} m^2) a result of 10^{126} STEPs are needed to duplicate the surface area needed for a correct cosmological constant (Λ). A marginal error of +/- 10^3 power can be explained by rounding or by broad approximations.

Therefore 10^{183} STEPs of the entire are reduced to +/- 10^{126} STEPs that are left in space-time after all of the matter is pulled out of the vacuum. These 10^{57} STEPs ~= 10^{60} STEPs relate to the # of shroud STEPs that contain all of the elementary matter particle Planck strings.

When the shroud steps are deducted from the total count, a total of +/- 10^{126} STEPs is needed to calculate for a true vacuum where matter is void from the true vacuum.

When only the surface area of the remaining STEPs ($\sim 10^{126+/-}$ STEPs) are used in the Ef equation, it also reduces to *[Ef=DmΛ$_a$]* **T, Θ,** *since* 10^{60} STEPs out of 10^{183} STEPs are not involved in a true vacuum, their surface area contributions are not needed to calculate the vacuum energy density.

In summery;

The Cosmological constant is calculated from the surface area of +/-10^{126} STEPs that create the true vacuum of space-time of the OU and the remaining 10^{60} STEPs are associated with normal matter as shrouds where Planck size string like particles reside.

At the dawn of time two singularities entangled and the Eternal force (Ef) was unleashed. The Ef settled as two alternating highly concentrated Dark Energies (Dark matters) that appeared as Zero Point Energy (ZPE) in a one cubic Planck's cube. This first embryonic cube is referred to as a specialized Space-Time-Energy-Particle (STEP) and by being specialized it is de-noted as a Space-Time-Energy-Nugget (STEN).

Some of the ZPE (~5%) was converted to normal matter (Nm) after a few Planck tic of time. This normal matter has an Einstein exchange with normal energy (Ne) by E= mc^2. Before inflation this normal matter relates to 10^{80} Neutrons and would be destroyed quickly if not for the notion that they are actually considered to be nucleus neutrons.

The NIST Center for Neutron Research says;

"While neutrons in stable nuclei can exist for an eternity, a free neutron hangs around for about 15 min. before it decays via the weak interaction to a proton, an electron and an antineutrino." (See Part IV)

Also as I indicated before: the 10^{80} neutrons, first created possess all of the elementary particles and when they expand during inflation with 10^{60} shroud STEPs, they are all intact. Only after inflation, will decay and collisions occur. This is referred to by others as the, electroweak epoch. This epoch period is similar to what the large hadron collider simulates except for the neutron decay. The 10^{80} original neutrons, with all their elementary string particles that have decayed and collided follow the same pathways as outlined by others.

The 15 min. when neutron decay occurs is more than enough time, when coupled with inflation, to solve the Horizon, Flatness and Magnetic-monopole Problems.

The reheating during this time is basically due to the collisions of neutrons, protons and elementary particles that spewed out during this high energy state.

Dark energy (De) has an exchange with Dark matter (Dm) where $De=DmC^2$, which will be explained latter, but first the original Dm has a phase shift. Some of the Dm (~5%) is converted to normal matter (Nm), by the equation $Dm=De/C^2 +NmC^2$. Dm/DE occupies 95% of the OU. Throughout the remainder of the OU expansion the Nm and Ne will remain constant at 5%, while the Dm converts to De and also remain constant at 95%. Only their respective ratios will change. (See Fig#26)

Therefore, $ZPE= Dm-Nm$, and by rearrangement $Dm=ZPE/Nm$ or $Dm=ZPE +nM$. Also $Dm=De/C^2$ when Nm is not involved.

I hypothesized that the OU expands by the conversion of Dm to De using the preceding equation $Dm=De/C^2$, or $De=DmC^2$.

It has been showed by Hubble that the expansion of the universe is accelerating now and just after inflation at an acceleration of Λ_a.

The four forces that spun from Ef are incorporated into the Holy Grail Equation by conjecture that some of the dark mass congeals into force

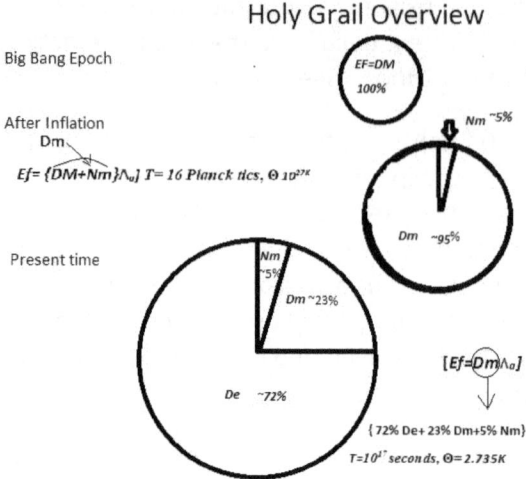

Holy Grail Overview

Big Bang Epoch

EF=DM
100%

After Inflation
Dm

$Ef= \{DM+Nm\}\Lambda_a]$ T= 16 Planck tics, Θ 10^{27K}

Nm ~5%

Dm ~95%

Present time

Nm ~5%

Dm ~23%

De ~72%

$[Ef=Dm\Lambda_a]$

{72% De+ 23% Dm+5% Nm}

$T=10^{1^*}$ seconds, $\Theta= 2.735K$

Fig#26. Pie graph of the Holy Grail overview.

string particles such as the bosons (Higgs, photon, gluon, graviton or weak particles such as the W+/- and the Z particle).

Fermions (mass particle strings) congeal also by this transformation and are incorporated into the Holy Grail Equation by Einstein's rearranged equation of $m=e/c^2$ from $e=mc^2$.

All Sparticles would be part of the Holy Grail equation by way of the M theory.

The eleven dimensions of space time were carved out during the early stages of STEP formation as the Ef was cooling.

Basic quantities and dimensions, that are derived from **$[Ef=Dm\Lambda_a]$ T, Θ** are listed in tables 3, 4 and 5 below.

Table 3. SI base units

Base quantity	Name	Symbol
length	meter	m
mass	kilogram	kg
time	second	s
electric current	ampere	A
thermodynamic temperature	kelvin	K
amount of substance	mole	mol
luminous intensity	candela	cd

Flatness Problem

Currently the Universe is not expanding so slowly that it will collapse back on itself in a Big Crunch. Also it is not expanding so quickly that it will keep expanding forever. Instead, results indicate that the Universe is trending toward a fine line between the two. The OU is referred to as "flat". Physicist did not expect this to be the case unless there were some cause for the OU to have such a fine line. Finding the cause is referred to as "the flatness problem".

I hypnotize that the OU will follow this trend ($Dm=De/c^2$), until the Dm depletes. This will be followed by Nm fusing by graviton forces and the STEPs will fold back into STENs which will cause the Big Crunch to prevail!

Table 4. Examples of SI derived units expressed in terms of SI base units

Derived quantity	SI derived unit	
	Name	Symbol
area	square meter	m^2
volume	cubic meter	m^3
speed, velocity	meter per second	m/s
acceleration	meter per second squared	m/s^2
wavenumber	reciprocal meter	m^{-1}
density, mass density	kilogram per cubic meter	kg/m^3
specific volume	cubic meter per kilogram	m^3/kg
current density	ampere per square meter	A/m^2
magnetic field strength	ampere per meter	A/m
luminance	candela per square meter	cd/m^2

When STEPs have excess De, they are detected as Dm by gravitational means only.

Table 5: Dimensional universal physical constants

normalized with Planck units

Constant	Symbol	Dimension	Value in SI units with uncertainties
Speed of light in vacuum	c	$L\ T^{-1}$	2.99792458×10^8 m s^{-1} (exact by definition of meter)
Gravitational constant	G	$L^3\ M^{-1}\ T^{-2}$	$6.67384(80)\times10^{-11}$ m^3 kg^{-1} s^{-2}
Reduced Planck constant	$\hbar = h/2\pi$ where h is Planck constant	$L^2\ M\ T^{-1}$	$1.054571726(47)\times10^{-34}$ Js
Coulomb constant	$(4\pi\varepsilon_0)^{-1}$ where ε_0 is the permittivity of free space	$L^3\ M\ T^{-2}$ Q^{-2}	$8.9875517873681764\times10^9$ kg m^3 s^{-2} C^{-2} (exact by definitions of ampere and meter)
Boltzmann constant	k_B	$L^2\ M\ T^{-2}$ Θ^{-1}	$1.3806488(13)\times10^{-23}$ J/K

Key: L = length, M = mass, T = time, Q = electric charge, Θ = temperature.

Tables 3,4,&5 are from Wickapedia.

Big bang epoch Conditions

When Tempature (Θ) is at 10^{32} Kalvin at (T) time= 10^0 Planck's tic just after the big bang, the equasion *[Ef=DmΛ_a]* T, Θ reduces to *[Ef=Dm]* T, Θ, since the OU is one cubic Planck's unit with no expansion.

Λ_a =10^0=1 Planck tic

At the beginning (the dawn of time and space) all of the Eternal force is manifested as Dark matter (concentrated Dark energy) *Dm=De/C^2, or De=*DmC^2.

163

Before inflation (the normal matter epoch)

The temperature cooled just enough to create 10^{80} normal matter neutrons compartmented in 10^{80} STENs from dark matter along with a sparse amount of De, $Dm=De/C^2 +NmC^2$. During this short period of time these 5% normal matter neutrons were confined in the STENs. The remaining 95% Dm/De lodged themselves as ZPM/ZPE exchanging geometry (and position) 10^{43} times a second. They are considered Calabi Yau tesseracts with one having a Planck cubic configuration on a 3-sphere. Dark energy (De) has an exchange with Dark matter (Dm) where $De=DmC^2$, which will be explained latter, but first the original Dm has a phase shift. Some of the Dm (~5%) is converted to normal matter (Nm), by the equation $Dm=De/C^2 +NmC^2$. Dm/DE occupies 95% of the OU. Throughout the remainder of the OU expansion the Nm and Ne will remain constant at 5%, while the Dm converts to De and also remain constant at 95%. Only their respective ratios will change. (See Fig#26)

Therefore, $ZPE= Dm-Nm$, and by rearrangement $Dm=ZPE/Nm$ or $Dm=ZPE +nM$. Also $Dm=De/C^2$ when Nm is not involved.

After the inflation epoch

The Temperature (Θ)cools to 10^{27} at (T) time 10^{-32} seconds and the equasion is $[Ef=Dm\wedge_a]$ $T=10^{16\ Planck\ tics}$ $\Theta =10^{\wedge}27K$, and $De= \{DM+Nm\}$ $C2$. Normal matter has congealed out of dark matter $Dm=De/C^2 +NmC^2 at$ this time and temperature.

Modern Universe

Temperature (Θ) =2.735 degrees above absolute zero and the time is 10^{17} seconds of our present OU.

Note:

When I discuss the surface area of a STEP, I am referring to the onion skin region between the two Calabi yau tesseract ZPEs. This region is also where the graviton operates, so it is not hard to imagine that this is the event horizon of a STEP.

Basic equations, quantities and dimensions all fit into the Holy Grail equation by normal substation of one unit for another. Earlier we substituted E for F and A for C when we went from F=ma to E=mc². We use this same technique in our Holy Grail Equation. For example, Λ can be substituted with its equal from Einstein's General relativity theory. Dm can be substituted with its energy equivalent of De/C² which in turn allows for Dark Energy calculations from Quantum theory.

The Holy Grail equation, *[Ef=DmΛₐ]* T, Θ contains every physics and mathematical relations there is in the entire OU. For example, if you wanted the cosmological constant equivalent in the equation, you would use the equation $Λ_a=ΛR/3$ for the acceleration of the OU and then substitute $ΛR/3$ into the Holy Grail equation *[Ef=Dm {ΛR/3}]* **T, Θ.**, where R=the size of the universe.

Furthermore, Einstein's field equation, from Wikipedia, (EFE) is usually written in the form:

$$R_{\mu\nu} - \frac{1}{2}Rg_{\mu\nu} + \Lambda g_{\mu\nu} = 8\pi\frac{G}{c^4}T_{\mu\nu}$$

Where

- $R_{\mu\nu}$ is the Ricci curvature tensor
- R is the Ricci scalar (the tensor contraction of the Ricci tensor)
- $g_{\mu\nu}$ is a (symmetric 4 x 4) metric tensor

- Λ is the Cosmological constant
- π is pi=3.1415..., the ratio between a circle's circumference and diameter
- G is the Gravitational constant
- c is the speed of light in free space
- $T_{\mu\nu}$ is the energy-momentum stress tensor of matter

This equation can be rearranged and converted to {Λ = 8π (G/c²) ρvac = κ ρ_{vac},} where κ is Einstein's constant}. It then can be solved for the cosmological constant and in turn be inserted into the Holy Grail equation.

Likewise, dark energy can be substituted for dark matter, etc. as seen earlier which allows for all quantum, string and other physics to be incorporated into the equation. This equation is the only expression that includes Dark matter/energy to be expressed mathematically with the entire OU.

The reason De is postulated to equal DmC² is because when you divide 10^{17} seconds of the OU, into the OU diameter 10^{27} meters you get 10^{10} ms⁻¹, which is only 2 magnitudes of order more than the speed of light. This discrepancy could be caused by rounding errors and or improper estimations of sizes etc.

This relationship of Dm and De indicates that Einstein's speed limit of 299,792,458 meters per second ≈3.00×10⁸ m/s or speed of light in mph (miles per hour, 670,616,629 mph) for matter and energy also applies to the expansion of the OU by the conversion of Dm to De.

Part VII:

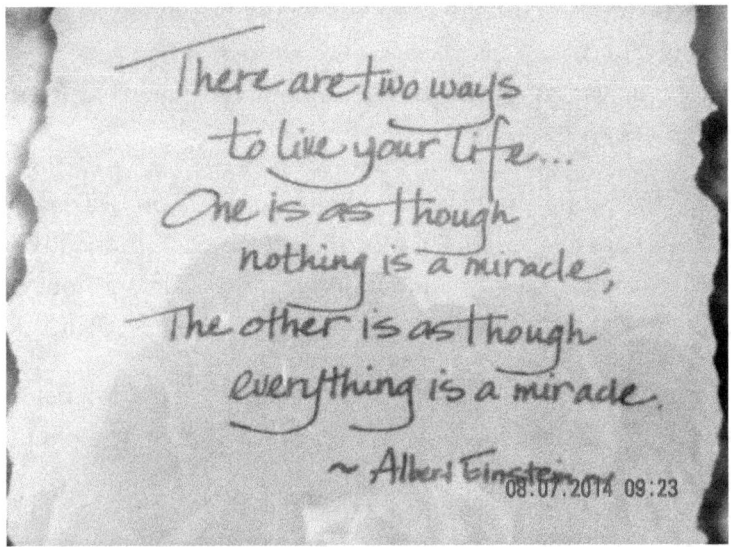

There are two ways
to live your life...
One is as though
nothing is a miracle,
The other is as though
everything is a miracle.
~ Albert Einstein

Eternity, Spirituality and our Universe

"Science without religion is lame. Religion without science is blind."

"I want to know how god created this world. I am not interested in this or that phenomenon, in the spectrum of this or that element. I want to know his thoughts; the rest are just details."

"My religion consists of a humble admiration of the illimitable superior spirit who reveals himself in the slight details we are able to perceive with our frail and feeble mind."

The Universe's relevance with Eternity

An Einstein quote sum's up best the quest for the thirst of knowledge.

" The important thing is not to stop questioning. Curiosity has its own reason for exiting. One cannot help be in awe when one contemplates the mysteries of eternity."

Earlier in this book I postulated that the universe was the direct result of a space-time-energy burst from two singularities that were part of an infinite Eternity. I also articulated that infinity concepts would be limited in scope for our observable universe. Planck's sizes minimums replace zero dimensions and 10^{500} universes replace the use of infinity when we refer to maximum limits.

Only pi and a few other concepts have an infinite connotation. However, none of these infinities are ever used. For example, pi is always rounded off to a finite number in any calculation and infinity is mostly factored out in others. Actually quantum theory would have pi end after the 10^{500} digit.

Also Physicist should avoid dividing or multiplying by "0" whenever possible. Using Planck units like 10^{-35} meters for "0" and 10^{500} for infinity when applicable would be useful.

At the forefront of my ideas about our universe, I indicated that the universe is finite and that most everything about it, like ultimate size, time, space- membranes, matter/energy ...are also finite.

I also indicated that our universe came from two singularity of an infinite Eternity where there is an infinite number of singularities that make up Eternity.

Thus far I have not considered what relationship life in our universe has to do with the singularities that produced our Universe!

By Life, I mean any living thing that habitat's our universe which I will also label as "Souls".

Souls are in no way a component derived from the singularity that produced our universe, but rather each Soul is equivalent to a singularity in themselves.

It's probably hard to think of a twig or a bacterium to have a soul, but keep in mind that life on earth, or elsewhere in outer space-time, is a temporary thing and to get hung up on soul capability is another subject in itself.

Here I will focus only on the Soul(s) of humans.

The Human Soul is an infinite entity and is as complex as the universe.

Our creator (who I and millions of others referred to as GOD) had a plan!

God's plan was and is a plan for Human Souls to make choices.

Currently, our observable universe has 10^{183} STEPs. Souls are just as complex but are only associated to the observable universe so as to make choices that may enhance or demote their existence.

Hitler made some very bad choices whereas Gandhi made some very good ones. We as humans have the same opportunity as they did in their life time.

What happened to them and what can happen to us after death is held in our own individual beliefs.

It is my belief, as a Christian and a human, that Karma is at play here and that after death is a continuation of life on earth. Only, this time (after death), without the restrictions of Quantum/STEP Reality but rather with Eternal Reality. All Souls are less restricted in the Eternity realm.

Eternal Reality comes with the passing of the Human Spirit to the Eternal Spirit along with the just acquired lifetime experience of human existence. This acquired experience either enhances or demotes the quality of being at this next stage of a Soul's infinite journey. Any history of a Souls journey can be found scripted in the Soul's DNA that, as of now, is not legible.

This view departs a little from Einstein's view.

Einstein, as a Jew and an independent thinker, did not have a personal God but rather his religious views as documented **by Wikipedia**, the free encyclopedia (6/1/2014) are as followers in italics with remarks from me as I see it.

Albert Einstein's religious views have been studied extensively. He said he believed in the "pantheistic" God of Baruch Spinoza, but not in a personal god, a belief he criticized. He also called himself an agnostic, while disassociating himself from the label atheist, preferring, he said, "an attitude of humility corresponding to the weakness of our intellectual understanding of nature and of our own being."

Einstein was raised by secular Jewish parents. In his Autobiographical Notes, Einstein wrote that he had gradually lost his faith early in childhood:

I came—though the child of entirely irreligious (Jewish) parents— to a deep religiousness, which, however, reached an abrupt end the age of twelve. Through the reading of popular scientific books I soon reached the conviction that much in the stories of the Bible could not be true. The consequence was a positively fanatic orgy of freethinking coupled with the impression that youth is intentionally being deceived by the state through lies; it was a crushing impression. Mistrust of every kind of authority grew out of this experience, a skeptical attitude toward the convictions that were alive in any specific social environment—an attitude that has never again left me, even though, later on, it has been tempered by a better insight into the causal connections. It is quite clear to me that the religious paradise of youth, which was thus lost, was a first attempt to free myself from the chains of the 'merely personal,' from an existence dominated by wishes, hopes, and primitive feelings.

Out yonder there was this huge world, which exists independently of us human beings and which stands before us like a great, eternal riddle, at least partially accessible to our inspection and thinking. The contemplation of this world beckoned as a liberation, and I soon noticed that many a man whom I had learned to esteem and to admire had found inner freedom and security in its pursuit.

The mental grasp of this extra-personal world within the frame of our capabilities presented itself to my mind, half consciously, half unconsciously, as a supreme goal. Similarly motivated men of the present and of the past, as well as the insights they had achieved, were the friends who could not be lost. The road to this paradise was not as comfortable and alluring as the road to the religious paradise; but it has shown itself reliable, and I have never regretted having chosen it.

I also developed some similar concepts about religion when I was around twelve. In 2008, I completed a fiction novel titled "Infinity of the Soul" in which I took the name of Kinen and referred to my mother as Ruth and told a true story about my youthful religious experience....

"Living on a farm several miles from any town, not having a car and living on welfare was very isolating to say the least. The school bus was basically the main means to connect to any type of sanity. One day, Kinen was working in the vegetable garden when he saw an old car with the trunk cut out and made into a human cattle car with wooden benches built where the trunk used to be. The driver was an old codger with a stinky, patched sports coat and old wash pants on. He parked along the side of the road and walked up the small hill to the front of the house and onto the front porch. He knocked on the door and went in when Kinen's mother greeted him. After about an hour, he left and drove back in the direction in which he came.

Later that evening, Kinen's mom, Ruth, told the kids that this man was from the local church, which was in the opposite direction of school. She said that he would pick them up on Sunday at 8:00 a.m. and bring them home after Sunday school and church service.

It was a good thing that the church was not in West Glenn, so at least his classmates wouldn't be there. The church was in Macade and only a few miles from the little white schoolhouse where his mother went to school when she younger.

Kinen was not a happy camper but the next Sunday all four of us kids were bundled up and carted off to church and Sunday school. This was a grueling experience. Sunday school was ok but the Holy Roller, brimstone, fire-eating minister was agonizing. On top of this, he solicited testimonials from the congregation. This was total torture and I was completely beside himself. I went home and cried to my mother, but my sisters liked being with the boys in church and thought church was great. They won out and church was a mainstay.

Sometimes, I would hide but would always wind up in the doghouse. This went on for months and eventually the church got my mother Ruth to attend. She would actually not only go in the morning but also go again on Sunday evening. She even made me participate at night for a few times until I absolutely refused to attend. This experience had such a negative effect that I resolved himself that I would go to hell when I died. In the back of my mind, I wondered if this whole business of religion was legit.

The outlandish minister preached that the only way you could get to heaven was to confess all your sins and never do them again. He also preached that any bad thoughts were sins. One time, before I stopped going, my mother dragged me along to a Sunday night prayer meeting. I saw my mother breaking down because of repeated pressure from the priest to go to the altar and be saved. She was crying and the church congregation was crying and I was motioned to come up to the altar and be saved too. Emotions were running rampant and I reluctantly went to the altar and knelt by my mother's side. I prayed to be forgiven for my sins.

The next couple weeks, I tried to be good, but as soon as I prayed to have God forgive my sins, I would have another bad thought so I would have to say another prayer to be forgiven for that one! This went on for several days. I noticed my mother didn't seem to be much different than before, so I asked her if she was saved. She told me that she felt that she was pressured into going up to the altar. It was easier to do that than resist. I didn't like this and I didn't like having to pray every half hour to have my last thought cleansed so I could still go to heaven! That is why I decided to just go to hell after death and not be manipulated by the church like my mother.

However, this was a compromise since I felt deep down that there must be life after death and that it should not be predicated on what you thought about church. I remembered that when I looked into my grandmother's mirror while holding another mirror and saw my image many, many times, I felt that I too could have many, many lives and that they should be based on my actions on Earth and not on whether I believed in a certain religion or, for that matter, a certain branch of a religion. I felt that my accomplishments as well as my shortcomings should somehow be balanced and rectified so that in another life I would be a product of my deeds."

As I described my experience as a youth in this book, I can see the parallel of my rational to Einstein's youthful rebellion.

Beliefs

Einstein used many labels to describe his religious views, including "agnostic", "religious nonbeliever" and a "pantheistic" believer in "Spinoza's God."

Personal God and the afterlife

Einstein expressed his skepticism regarding an anthropomorphic deity, often describing it as "naïve" and "childlike". He stated, "It seems to me that the idea of a personal God is an anthropological concept which I cannot take seriously. I feel also not able to imagine some will or goal outside the human sphere. My views are near those of Spinoza: admiration for the beauty of and belief in the logical simplicity of the order which we can grasp humbly and only imperfectly. I believe that we have to content ourselves with our imperfect knowledge and understanding and treat values and moral obligations as a purely human problem—the most important of all human problems."

Einstein may be right about moral obligations being only a human problem since Paul Zak experiments show the presence of a chemical, oxytocin that makes people more willing to help a stranger. Those who release the most oxytocin is the most satisfied with their lives. He says Aristotle was right in saying that the reason to be virtuous is that it makes us happy.

However, OxyContin is matter so indirectly morality stems from the creation of the universe. This is where I disagree with Einstein since I believe that the universe was created for "Souls" to habitat and make

173

decisions which will have consequences when returning to Eternity. Einstein felt that one life was enough, but when he was living he was not privy to the current knowledge (theory) of STEPs.

On 22 March 1954 Einstein received a letter from Joseph Dispentiere, an Italian immigrant who had worked as an experimental machinist in New Jersey. Dispentiere had declared himself an atheist and was disappointed by a news report which had cast Einstein as conventionally religious. Einstein replied on 24 March 1954:

It was, of course, a lie what you read about my religious convictions, a lie which is being systematically repeated. I do not believe in a personal God and I have never denied this but have expressed it clearly. If something is in me which can be called religious then it is the unbounded admiration for the structure of the world so far as our science can reveal it.In a letter to Beatrice Frohlich, 17

December 1952 Einstein stated, "The idea of a personal God is quite alien to me and seems even naïve." Eric Gutkind sent a copy of his book "Choose Life: The Biblical Call To Revolt" to Einstein in 1954. Einstein sent Gutkind a letter in response and wrote, "The word God is for me nothing more than the expression and product of human weaknesses, the Bible a collection of honourable, but still primitive legends. No interpretation no matter how subtle can (for me) change this. These subtilised interpretations are highly manifold according to their nature and have almost nothing to do with the original text."On 24 April 1929, Einstein cabled Rabbi Herbert S. Goldstein in German: "I believe in Spinoza's God, who reveals himself in the harmony of all that exists, not in a God who concerns himself with the fate and the doings of mankind." He expanded on this in written answers he gave to a Japanese scholar on his views on science and religion, which appeared as a limited edition publication, on the occasion of Einstein's 50th birthday: Scientific research can reduce superstition by encouraging people to think and view things in terms of cause and effect. Certain it is that a conviction, akin to religious feeling, of the rationality and intelligibility of the world lies behind all scientific work of a higher order... This firm belief, a belief bound up with a deep feeling, in a superior mind that reveals itself in the world of experience, represents my conception of God. In common parlance this may be described as "pantheistic" (Spinoza).

On the question of an afterlife Einstein stated to a Baptist pastor, "I do not believe in immortality of the individual, and I consider ethics to be an exclusively human concern with no superhuman authority behind it. "This sentiment was also expressed in Einstein's The World as I See It, stating: "I cannot conceive of a God who rewards and punishes his creatures, or has a will of the type of which we are conscious in ourselves. An individual who should survive his physical death is also beyond my comprehension, nor do I wish it otherwise; such notions are for the fears or absurd egoism of feeble souls. Enough for me the mystery of the eternity of life, and the inkling of the marvellous structure of reality, together with the single-hearted endeavour to comprehend a portion, be it never so tiny, of the reason that manifests itself in nature."

Agnosticism, Deism and Atheism

Einstein rejected the label atheist. Einstein stated: "I have repeatedly said that in my opinion the idea of a personal God is a childlike one. You may call me an agnostic, but I do not share the crusading spirit of the professional atheist whose fervor is mostly due to a painful act of liberation from the fetters of religious indoctrination received in youth. I prefer an attitude of humility corresponding to the weakness of our intellectual understanding of nature and of our own being. "According to Prince Hubertus, Einstein said, "In view of such harmony in the cosmos which I, with my limited human mind, am able to recognize, there are yet people who say there is no God. But what really makes me angry is that they quote me for the support of such views."

Einstein had previously explored the belief that man could not understand the nature of God. In an interview published in 1930 in

G. S. Viereck's' book Glimpses of the Great, Einstein, in response to a question about whether or not he believed in God, explained:

Your question [about God] is the most difficult in the world. It is not a question I can answer simply with yes or no. I am not an Atheist.

I do not know if I can define myself as a Pantheist. The problem involved is too vast for our limited minds. May I not reply with a parable? The human mind, no matter how highly trained, cannot grasp the universe. We are in the position of a little child, entering a huge library whose walls are covered to the ceiling with books in many different tongues. The child knows that someone must have written those books. It does not know who or how. It does not understand the languages in which they are written. The child notes a definite plan in the arrangement of the books, a mysterious order, which it does not comprehend, but only dimly suspects. That, it seems to me, is the attitude of the human mind, even the greatest and most cultured, toward God. We see a universe marvelously arranged, obeying certain laws, but we understand the laws only dimly. Our limited minds cannot grasp the mysterious force that sways the constellations. I am fascinated by Spinoza's Pantheism. I admire even more his contributions to modern thought. Spinoza is the greatest of modern philosophers, because he is the first philosopher who deals with the soul and the body as one, not as two separate things.

In a 1950 letter to M. Berkowitz, Einstein stated that "My position concerning God is that of an agnostic. I am convinced that a vivid consciousness of the primary importance of moral principles for the betterment and ennoblement of life does not need the idea of a law-giver, especially a law-giver who works on the basis of reward and punishment."

According to biographer Walter Isaacson, Einstein was more inclined to denigrate disbelievers than the faithful. Einstein said in correspondence, "[T]he fanatical atheists...are like slaves who are still feeling the weight of their chains which they have thrown off after hard struggle. They are creatures who—in their grudge against the traditional 'opium of the people'—cannot bear the music of the spheres." Although he did not believe in a personal God, he indicated that he would never seek to combat such belief because "such a belief seems to me preferable to the lack of any transcendental outlook."

In 1945 Guy Raner, Jr. wrote a letter to Einstein, asking him if it was true that a Jesuit priest had caused Einstein to convert from atheism.

Einstein replied, "I have never talked to a Jesuit priest in my life and I am astonished by the audacity to tell such lies about me. From the viewpoint of a Jesuit priest I am, of course, and have always been an atheist. ... It is always misleading to use anthropomorphical concepts in dealing with things outside the human sphere—childish analogies. We have to admire in humility the beautiful harmony of the structure of this world—as far as we can grasp it, and that is all."

Determinism

Like Spinoza, Einstein was a strict determinist who believed that human behavior was completely determined by causal laws. For that reason, he refused the chance aspect of quantum theory, rejecting the concept of a god playing dice with the universe. In letters sent to physicist Max Born, Einstein revealed his devout belief in causal relationships:

You believe in a God who plays dice, and I in complete law and order in a world which objectively exists, and which I in a wildly speculative way, am trying to capture. I firmly believe, but I hope that someone will discover a more realistic way, or rather a more tangible basis than it has been my lot to find. Even the great initial success of the quantum theory does not make me believe in the fundamental dice game, although I am well aware that some of our younger colleagues interpret this as a consequence of senility.

STEP synchronization with quantum probabilities is the answer!

Einstein's emphasis on 'belief' and how it connected with determinism was illustrated in a letter of condolence responding to news of the death of one of his lifelong friends. Einstein wrote to the family: "Now he has departed from this strange world a little ahead of me. That signifies nothing. For us believing physicists the distinction between past, present, and future is only a stubbornly persistent illusion."

Einstein had admitted to a fascination with philosopher Spinoza's deterministic version of pantheism. American philosopher Charles

Hartshorne, in seeking to distinguish deterministic views with his own belief of free will panentheism, coined the distinct typology "Classical pantheism" to distinguish the views of those who hold similar positions to Spinoza's deterministic version of pantheism.

Moral philosophy

Einstein was a Humanist and a supporter of the Ethical Culture movement. He served on the advisory board of the First Humanist Society of New York. For the seventy-fifth anniversary of the New York Society for Ethical Culture, he stated that the idea of Ethical Culture embodied his personal conception of what is most valuable and enduring in religious idealism. He observed, "Without 'ethical culture' there is no salvation for humanity." He was an honorary associate of the British Humanist organization, the Rationalist Press Association and its journal was among the items present on his desk at his death.

With regard to Divine command theory, Einstein stated, "I cannot imagine a God who rewards and punishes the objects of his creation, whose purposes are modeled after our own—a God, in short, who is but a reflection of human frailty. Neither can I believe that the individual survives the death of his body, although feeble souls harbor such thoughts through fear or ridiculous egotisms. A God who rewards and punishes is inconceivable to him for the simple reason that a man's actions are determined by necessity, external and internal, so that in God's eyes he cannot be responsible, any more than an inanimate object is responsible for the motions it undergoes. Science has therefore been charged with undermining morality, but the charge is unjust. A man's ethical behavior should be based effectually on sympathy, education, and social ties and needs; no religious basis is necessary. Man would indeed be in a poor way if he had to be restrained by fear of punishment and hopes of reward after death. It is therefore easy to see why the churches have always fought science and persecuted its devotees."

On the importance of ethics he wrote, "The most important human endeavor is the striving for morality in our actions. Our inner balance and even our very existence depend on it. Only morality in our actions can give *beauty and dignity to life. To make this a living force and bring it to clear consciousness is perhaps the foremost task of education. The foundation of morality should not be made dependent on myth nor tied to any authority lest doubt about the myth or about the legitimacy of the authority imperil the foundation of sound judgment and action." "I do not believe that a man should be restrained in his daily actions by being afraid of punishment after death or that he should do things only because in this way he will be rewarded after he dies. This does not make sense. The proper guidance during the life of a man should be the weight that he puts upon ethics and the amount of consideration that he has for others." I cannot conceive of a personal God who would directly influence the actions of individuals, or would directly sit in judgment on creatures of his own creation. I cannot do this in spite of the fact that mechanistic causality has, to a certain extent, been placed in doubt by modern science. My religiosity consists in a humble admiration of the infinitely superior spirit that reveals itself in the little that we, with our weak and transitory understanding, can comprehend of reality. Morality is of the highest importance—but for us, not for God."*

Cosmic spirituality

In his 1949 book The World as I See It, he wrote: "A knowledge of the existence of something we cannot penetrate, of the manifestations of the profoundest reason and the most radiant beauty, which are only accessible to our reason in their most elementary forms—it is this knowledge and this emotion that constitute the truly religious attitude; in this sense, and in this alone, I am a deeply religious man."

Einstein referred to his belief system as "cosmic religion" and authored an eponymous article on the subject in 1954, which later became his book Ideas and Opinions in 1955. The belief system recognized a "miraculous order which manifests itself in all of nature as well as in the world of ideas," devoid of a personal God who rewards and punishes individuals based on their behavior. It rejected a conflict between science and religion, and held that cosmic religion was necessary for science. He told William Hermanns in an interview that "God is a mystery. But a comprehensible mystery. I have nothing but awe when I observe the laws of nature. There are not laws without a lawgiver, but how does this lawgiver look? Certainly not like a man magnified." He added with a smile "some centuries ago I would have been burned or hanged. Nonetheless, I would have been in good company."

In a 1930 New York Times article, Einstein distinguished three human impulses which develop religious belief: fear, social morality, and a cosmic religious feeling. A primitive understanding of causality causes fear, and the fearful invent supernatural beings analogous to themselves. The desire for love and support create a social and moral need for a supreme being; both these styles have an anthropomorphic concept of God. The third style, which Einstein deemed most mature, originates in a deep sense of awe and mystery. He said, the individual feels "the sublimity and marvelous order which reveal themselves in nature ... and he wants to experience the universe as a single significant whole." Einstein saw science as an antagonist of the first two styles of religious belief, but as a partner in the third. He maintained, "even though the realms of religion and science in themselves are clearly marked off from each other" there are "strong reciprocal relationships and dependencies" as aspirations for truth derive from the religious sphere. For Einstein, "science without religion is lame, religion without science is blind."

He continued: a person who is religiously enlightened appears to me to be one who has, to the best of his ability, liberated himself from the fetters of his selfish desires and is preoccupied with thoughts, feelings and aspirations to which he clings because of their super-personal value. It seems to me that what is important is the force of this super personal content ... regardless of whether any attempt is made to unite this content with a Divine Being, for otherwise it would not be possible to count Buddha and Spinoza as religious personalities. Accordingly a religious person is devout in the sense that he has no doubt of the significance of those super-personal objects and goals which neither require nor are capable of rational foundation ... In this sense religion is the age-old endeavor of mankind to become clearly and completely conscious of these values and goals and constantly to strengthen and extend their effect. If one conceives of religion and science according to these definitions then a conflict between them *appears impossible. For science can only ascertain what is, but not what should be...*

An understanding of causality was fundamental to Einstein's ethical beliefs. In Einstein's view, "the doctrine of a personal God interfering with natural events could never be refuted, in the real sense, by science," for religion can always take refuge in areas that science can not yet explain. It was Einstein's belief that in the "struggle for the ethical good, teachers of religion must have the stature to give up the doctrine of a personal God, that is, give up that source of fear and hope" and cultivate the "Good, the True, and the Beautiful in humanity itself."

Jewish identity

In a letter to Eric Gutkind dated 3 January 1954, Einstein wrote in German, "For me the unaltered Jewish religion like all other religions is an incarnation of the most primitive superstitions. And the Jewish people to whom I gladly belong and with whose mentality I have a deep affinity have no different quality for me than all other people. As far as my experience goes, they are also no better than other human groups, although they are protected from the worst cancers by a lack of power. Otherwise I cannot see anything 'chosen' about them."

In an interview published by Time magazine with George Sylvester Viereck, Einstein spoke of his feelings about Christianity. Born in Germany in 1884 Viereck supported German nationalism, but was not anti-semitic. He was jailed in America in 1942 for being a German propagandist. Like Einstein Varieck was a pacifist; Varieck was accused of treason and expelled from the American Author's League because of his articles attacking war. At the time of the interview Einstein was informed that Viereck was not Jewish, but stated that Viereck had "..the psychic adaptability of the Jew," making it possible for Einstein to talk to him "without barrier." Viereck began by asking Einstein if he considered himself a German or a Jew, to which Einstein responded, "It's possible to be both." Viereck moved along in the interview to ask Einstein if Jews should try to assimilate, to which Einstein replied "We Jews have been too eager to sacrifice our idiosyncrasies in order to conform." Einstein was then asked to what extent he was influenced by Christianity.

"As a child I received instruction both in the Bible and in the Talmud. I am a Jew, but I am enthralled by the luminous figure of the Nazarene." Einstein was then asked if he accepted the historical existence of Jesus, to which he replied, "Unquestionably! No one can read the Gospels without feeling the actual presence of Jesus. His personality pulsates in every word. No myth is filled with such life."

He stressed however in a conversation with William Hermanns that, "I seriously doubt that Jesus himself said that he was God, for he was too much a Jew to violate that great commandment: Hear O Israel, the Eternal is our God and He is one!' and not two or three." Einstein lamented, "Sometimes I think it would have been better if Jesus had never lived. No name was so abused for the sake of power!" Nevertheless, he also expressed his belief that "if one purges the Judaism of the Prophets and Christianity as Jesus Christ taught it of all subsequent additions, especially those of the priests, one is left with a teaching which is capable of curing all the social ills of humanity."

Christian churches

The only Jewish school in Munich had been closed in 1872 for want of students, and in the absence of an alternative Einstein attended Catholic elementary school. He also received Jewish religious education at home, but he did not see a cleave between the two faiths, as he perceived the "sameness of all religions. "Einstein was equally impressed by the stories of the Hebrew Bible and the Passion of Jesus. According to biographer Walter Isaacson, Einstein immensely enjoyed the Catholic religion courses which he received at the school. The teachers at his school were liberal and generally made no distinction between student's religions, although some harbored an innate but mild antisemitism. Einstein later recalled an incident involving a teacher who particularly liked him, "One day that teacher brought a long nail to the lesson and told the students that with such nails Christ had been nailed to the Cross by the Jews" and that "Among the children at the elementary school anti-Semitism was prevalent...Physical attacks and *insults on the way home from school were frequent, but for the most part not too vicious." Einstein noted, "That was at a Catholic school; how much worse the antisemitism must be in other Prussian schools, one can only imagine." He would later in life recall that "The religion of the fathers, as I encountered it in Munich during religious instruction and in the synagogue, repelled rather than attracted me."*

In 1940 Time magazine quoted Einstein lauding the Church for its role in opposing the Nazis:

Only the Church stood squarely across the path of Hitler's campaign for suppressing truth. I never had any special interest in the Church before, but now I feel a great affection and admiration because the Church alone has had the courage and persistence to stand for intellectual truth and moral freedom. I am forced thus to confess that what I once despised I now praise unreservedly.

The quotation has since been repeatedly cited by defenders of Pope Pius XII. An investigation of the quotation by mathematician

William C. Waterhouse and Barbara Wolff of the Einstein Archives in Jerusalem found that the statement was mentioned in an unpublished letter from 1947. In the letter to Count Montgelas, Einstein explained that the original comment was a casual one made to a journalist regarding the support of "a few churchmen" for individual rights and intellectual freedom during the early rule of Hitler and that, according to Einstein, the comment had been drastically exaggerated.

In 2008 the Antiques Roadshow television program aired a manuscript expert, Catherine Williamson, authenticating a letter from Einstein in which he confirms that he "made a statement which corresponds approximately" to Time magazine's quotation of him, however "I made this statement during the first years of the Nazi regime—much earlier than 1940—and my expressions were a little more moderate."

On 11 November 1950 the Rev. Cornelius Greenway of Brooklyn wrote a letter to Einstein which had also quoted his alleged remarks about the

Church. Einstein responded, "I am, however, a little embarrassed. The wording of the statement you have quoted is not my own. Shortly after Hitler came to power in Germany I had an oral conversation with a newspaper man about these matters. Since then my remarks have been elaborated and exaggerated nearly beyond recognition. I cannot in good conscience write down the statement you sent me as my own. The matter is all the more embarrassing to me because I, like yourself, I am predominantly critical concerning the activities, and especially the political activities, through history of the official clergy. Thus, my former statement, even if reduced to my actual words (which I do not remember in detail) gives a wrong impression of my general attitude."

Catholic Cardinal William Henry O'Connell spoke about Einstein's perceived lack of belief, "The outcome of this doubt and befogged speculation about time and space is a cloak beneath which hides the ghastly apparition of atheism." A Bronx Rabbi criticized both the Cardinal and Einstein for opining on matters outside their expertise: "Einstein would have done better had he not proclaimed his nonbelief in a God who is concerned with fates and actions of individuals. Both have handed down dicta outside their jurisdiction. The Catholic priest and broadcaster Fulton Sheen—whose intellect Einstein admired, even calling him "one of the most intelligent people in today's world described Einstein's New York Times article "the sheerest kind of stupidity and nonsense."William Hermanns conversations.

Einstein's conversations with William Hermanns were recordedover a 34-year correspondence. In the conversations Einstein makes various statements about the Christian Churches in general and the Catholic Church in particular: "When you learn the history of the Catholic Church, you wouldn't trust the Center Party. Hasn't Hitler promised to smash the Bolsheviks in Russia? The Church will bless its Catholic soldiers to march alongside the Nazis" (March 1930). "I predict that the Vatican will support Hitler if he comes to power.

The Church since Constantine has always favoured the authoritarian State, as long as the State allows the Church to baptize and instruct the masses" (March 1930). "So often in history the Jews have been the instigators of

184

justice and reform whether in Spain, Germany or Russia. But no sooner have they done their job than their 'friends', often blessed by the Church, spit in their faces" (August 1943).

"But what makes me shudder is that the Catholic Church is silent. One doesn't need to be a prophet to say, 'The Catholic Church will pay for this silence...I do not say that the unspeakable crimes of the Church for 2,000 years had always the blessing of the Vatican, but it vaccinated its believers with the idea: We have the true God, and the Jews have crucified Him.' The Church sowed hate instead of love, though the Ten Commandments state: Thou shalt not kill" (August 1943). "With a few exceptions, the Roman Catholic Church has stressed the value of dogma and ritual, conveying the idea theirs is the only way to reach heaven. I don't need to go to Church to hear if I'm good or bad; my heart tells me this" (August 1943). "I don't like to implant in youth the Church's doctrine of a personal God, because that Church has behaved so inhumanly in the past 2,000 years...

Consider the hate the Church manifested against the Jews and then against the Muslims, the Crusades with their crimes, the burning stakes of the inquisition,(and) the tacit consent of Hitler's actions while the Jews and the Poles dug their own graves and were slaughtered. And Hitler is said to have been an altar boy!" (August 1943).

"Yes" Einstein replied vehemently, "It is indeed human, as proved by Cardinal Pacelli (the future Pope Pius XII), who was behind the Concordat with Hitler. Since when can one make a pact with Christ and Satan at the same time?" (August 1943). "The Church has always sold itself to those in power, and agreed to any bargain in return for immunity." (August 1943) "If I were allowed to give advice to the Churches," Einstein continued, "I would tell them to begin with a conversion among themselves, and to stop playing power politics. Consider what mass misery they have produced in Spain, South America and Russia." (September 1948).

In response to a Catholic convert who asked "Didn't you state that the Church was the only opponent of Communism?" Einstein replied, "I don't have to emphasise that the Church at last became a strong opponent of National Socialism, as well." Einstein's secretary Helen Dukas added, "Dr.

Einstein didn't mean only the Catholic church, but all churches." When the convert mentioned that family members had been gassed by the Nazis, Einstein replied that "he also felt guilty—adding that the whole Church, beginning with the Vatican, should feel guilt." (September 1948)

"About God, I cannot accept any concept based on the authority of the Church... As long as I can remember. I have resented mass indoctrination. I cannot prove to you there is no personal God, but if I were to speak of him, I would be a liar. I do not believe in the God of theology who rewards good and punishes evil. His universe is not ruled by wishful thinking, but by immutable laws" (1954). William Miller of Life Magazine who was present at this meeting described Einstein as looking like a "living saint" and speaking with "angelic indifference."

I can identify strongly with Einstein's view of Church and organized religion. It is my view that Einstein was more complacent about formal religion than he was about God. These were his beliefs and he made sure people didn't put words in his mouth.

The fact that he leaned toward pantheism and Spinoza and referred to his beliefs as a cosmic view and indicated that he has an agnostic point of view all points to the fact that he holds God in a very high place. I consider myself as an agnostic theism. Agnostic theism is the philosophical view that encompasses both theism and agnosticism. An agnostic theist believes in the existence of at least one deity, but regards the basis of this proposition as unknown or inherently unknowable.

My deity relates to Christianity. If Einstein was an agnostic theism his deity would most likely be Spinoza relating to the entire universe as his deity.

Einstein was certainly not a religious man in the sense that he would not take scripture literally and he apparently was not interested in the non-literal interpretation either. He professed that he did not believe in life after death but he also confessed that *"The human mind, no matter how highly trained, cannot grasp the universe.*

We are in the position of a little child, entering a huge library whose walls are covered to the ceiling with books in many different tongues."

His thought about "lack of knowledge" might be the reason he felt after death activity is a non-event.

Einstein, like Spinoza felt that the soul and the body are one, not as two separate things. If he saw Eternity and Souls as infinite, as I do, and the universe as finite then he may have reconsidered morality and after death in a different light, but still have some issues with organized religion.

But then again, who am I to put words in his mouth... or did he put words in mine? Bill O'Reilly from Fox Channel, "The Factor," clearly pointed to a divine intervention when he got inspiration to write his book "The Killing of Jesus".

Like Bill, I also would wake up in the middle of the night and record my thoughts from a sound sleep about the grand original design of our Universe, Eternity and Souls.

"The further the spiritual evolution of mankind advances, the more certain it seems to me that the path to genuine religiosity does not lie through the fear of life, and the fear of death, and blind faith, but through striving for rational knowledge."

The New Reality can only point to a universe that is "NOT" randomly sparked into existence, but rather put together by a Grand Original Design (G.O.D.).

Since all of the ~10^{183} STEPs are synchronized, the universe cannot be random because every event in the G.O.D. matrixes is a choice and not a statistical probability as in a quantum wave equation.

This is supported by the double split experiment and the "Spooky action at a distance" phenomena. In both cases matter follows the rules of a complete Quantum theory engulfing STEPs that use superluminal information to communicate throughout the universe by time travel and not by space-time travel, which would violate Einstein's speed limit for the speed of electrometric waves (i.e. the speed of light).

The New Reality eliminates atheistic concepts of the universe because there would not be enough random event to allow synchronization to take place.

Creativity trumps spontaneity!

Spontaneity would not allow for STEP synchronization!

Part VIII

Concepts in terms of definitions (a special glossary of terms)

"The whole of science is nothing more than a refinement of everyday thinking."

"Education is what remains after one has forgotten everything he learned in school."

Big Bang: The entanglement of two singularities that resulted in an extremely hot Space-Time-Energy-Nugget (STEN) housing Zero- Point-Energy(ZPE), during the first Planck's tic of time, in a Planck size cube. The ZPE, housed in this nugget as two alternating Dark forces (collectively the Eternal force) created great pressure on the space membrane walls of the nugget. The second tic enabled an onion-skin to develop between the two tesseracts, of the ZPE, by congealing some of the ZPE into a Graviton which has the potential to pull the STEN back to its origin, upon cooling. (note: this is the first "separation of forces"). The 3rd tic through 10^{-38} sec. resulted in all the element particles of the universe to congeal out of one tesseract (the tesseract on a three sphere). The particle conversion of ZPE follows Einstein-like equation $E=MC^2$. The particles emerged as neutron STENs, 10^{80} of them, each with 248 elementary particles (strings) compartmentalized as what Lisi Garrett calls "the high dimensional object." (Note: Garrett refers to the strings as points). The strong force emerged (2nd separation of force) and inflation took over. Inflation resulted from leaking of ZPE from the 10^{80} STEN's, only this leaking creates STEPs. The universe's diameter went from a Planck size nugget to about the size of the Panama Canal. The weak and electromagnetic force separated out and the Higgs field/Higgs particle was created at this time.

Electron/Positron: From the string theory all elementary particles are one Planck's length in size. An Electron is said to occupy a volume of $\sim 10^{-45}$ cubic meters, which $\sim 10^{60}$ STEPs. Thus the string-like electron must occupy

189

this volume as a standing wave. Dr. Frank Close calls this volume the electron shroud and hypothesizes that space energy pops in and out of reality within this shroud and this is where the electron gets its EMF from. I would add to this and project that only one of the ~10^{60} STEPs surrounding the electron string is activated at a time and that one is the one that the string is passing through at the time of energy activity. In a single second the STEP exchange energy-tesseract places every 10^{43} times, so the electron-matter-string practically moves through the entire 10^{60} STEPs in this time frame. However, when it moves through any of the entire ~10^{183} STEPs, it moves as a traveling wave with physical particle implications. As it moves along it keeps it's Shroud in tack which accounts for its 10^{-45} meter volume. The positron is the electron's antiparticle and has similar geometry to the electron except that it has a positive charge.

Dark matter: Is the Zero Point Energy that has yet to convert to Dark Energy which expands the observable universe (see Matter, dark).

Energy (Dark): Is associated to dark matter by the equation $De = DmC^2$, where De is dark energy, Dm is dark matter, C is the speed of light. De is found in STEPs at a level below the threshold of graviton attraction.

Energy, Normal, Ne: Is associated with normal matter by Einstein's equation $E = MC^2$.

Eternity: An infinite number of singularities.

Extraordinary Sensory Reception (ESP): Instantaneous alignment of STEPs often occurring when a catastrophic event occurs, so that what is known at the event is simultaneously know by the receptor. Information travels super- luminal (instantaneous) through time canals, but not through space-time so Einstein's speed limit is not violated.

Genius: An individual that has an average to above average knowledge base for his age but possess an extraordinary entangled creative ability. Albert Einstein was one such person.

Gravitational waves: When inflation occurred the "0" ring gravitons resisted (they wanted to pull the STENs back to the singularity but instead were stressed to the point where polarized waves were created), thus gravitational waves were produced that support the inflation theory.

Heisenberg's' Uncertainty Principle: the more precisely the position is known the more uncertain the momentum is and vice versa. Keep in mind that the key operating word here is "known"! Since we do not have the tool to measure location and distance/ time at the Planck's scale it does make Heisenberg's' Uncertainty Principle correct. But if we had calibrated Planck size tools my certainty/uncertainty principal would be valid.

Stauffer's Certainty and Uncertainty Principle: With Planck size tools, Planck size strings and brains are always measurable along with their momentum as long as they do not shrink to sizes less than a Planck unit or move faster than a Planck length/Planck tic of time (which would violate Einstein's speed limit in our universe).

Hydrogen Ion/Atom: The beta decay of the neutron into a proton and an electron etc. gave a negative EMF to the electron and a positive EMF force to the proton that permeated their respective volumes as energy shrouds. When this beta decay happened just after the big bang, the electron was flung outward in a very hostel heightened temperature environment. However, upon cooling some electrons found proton mates to equilibrate their charges is such a way as to result in hydrogen ion. Because STEPs are synchronized and have specific worm holes of travel at the micro and macro level, the electron captured by the proton has a confined volume to operate in while in the presence of its proton capture. It has a 95% chance of being in the S1 orbital of the hydrogen ion, and when two hydrogen ions come in proximity to each other a hydrogen atom is formed. Also, there is a 1 in 1,000 chance that a stray neutron will find its way to the mix. This results in a hydrogen atom isomer to form. But the main point I want to make here is that the S1 orbital that is formed is because the STEPs that are in the S1 orbital are synchronized just like a select set of gears in a wind up clock (i.e.) like the gears for the second hand) and maintain a specific volume. It has been reported that the nearest electron to the nucleus of an atom is analogous to the distance of a pin head at the center of the 50-yard line of a football field to a person in the upper deck of the stadium. The ratio of the nucleus radius to the nearest electron distance is therefore 1: 300,000.

Inflation: Is the expansion of the early universe, after string-matter was formed, by the rapid synthesis of additional Space-Time- Energy-Particles (STEPs).

Kinetic energy: Energy that is seen when string matter moves as a standing wave.

M theory: A string theory that has 11 dimensions and shows super-symmetry.

Matter (normal): Is congealed Dark Energy/Dark Matter that travels throughout our universe by way of the + ½ twist worm holes of the Space-Time-Energy-Particles (STEPs). All matter is vibrating Planck size strings (fermions).

Matter (Dark): Is abundant dark energy contained in specialized STEPs referred to as STENs. This concentrated dark matter converts to dark energy by $Dm=De/C^2$ power, where De is dark energy, Dm is dark matter, C is the speed of light. The concentration of dark energy (alias dark matter) is above the threshold required to attract gravitons, thus this allows gravitons to exert its force upon dark matter and be detected in our observable universe.

Neutron: A neutron occupies~10^{-43} cubic meters and would need ~10^{63} STEPs to make up this volume. As discussed earlier I hypothesize that Garrett's theory of 248 elementary point particles which he refers to as "the high dimensional object" is actually a neutron, only the point particles are strings. Currently the majority of the strings are dormant. Only the up, down quarks and gluons are active (vibrate) in the neutron. Only when neutrons are smashed together will they expose their content.

Particle Zoo: Are the 248 elementary particles proposed by Lisi Garrett as point particles but here are refer to as Planck size strings.

Proton: It has the same approximate size 10^{-43} cubic meters of a neutron but with a different profile of up and down quarks. It also lacks an electron and an anti-neutrino.

Singularities: Found in the infinite eternity and are an unknown entity which are smaller than a Planck unit. They are impossible to describe by our mathematical system.

Space membrane: Two dimensional Planck square M theory membranes (d-branes) which house Zero Point energy of tesseract-Calabi-Yau manifolds.

Space-Time-Energy-Nugget (STEN): A special type of space- time-energy-particle (STEP) that was created at the dawn of time and spawned 10^{80} more before inflation set in which then produced STEPs that sent the STENs in all directions. This separation of matter made the universe homogeneous and answered the flatness and horizon problem. STEN is a special type of STEP that has an abundance of Dark Energy/Matter.

Space-Time- Energy-Particle (STEP): ~10^{183} currently make up the observable universe. They have 11 dimensions. Each has the potential to be a universe. If quantum predictions are correct, then 10^{500} may be their limit number.

Standing wave: A wave that moves back on its self but can also travel throughout the observable universe. An example is the shroud around an electron.

Time: Measured over a Planck unit length allows Zero-point energy (ZPE) to exchange locations every 10^{-43} sec.

Traveling wave: A wave that travels through the observable universe with or without standing wave motion. An example is an electron, with its shroud traveling from point A to point B in an electronic device.

Volume: = 4D +7 additional dimensions.

= the region between space membranes that is accounted for by the passage of time through a distance = to the height & depth of the space (one Planck's length each) membrane divided by a one Planck's tic of time. The seven additional dimensions are two ½ twist wormholes, two 1 twist wormholes, one axle wormhole with a zero twist, one ringworm with a 3/2 twist and an onion skin region that has a 2 twist.

Warped Space: Is the rotation of synchronize Space-Time-Energy- Particles when gravitons pull string matter in an apparent curve-like manor (actually it might be a Planck length step-like directional change).

Zero-point Energy: Is the amount of dark energy that is left in a vacuum, the energy that remains when all other (normal) energy is removed from a system. It is also referred to as the energy in a vacuum at absolute zero degrees' kelvin. I postulate that it is Dark Energy that is equal to the positive cosmological constant of Einstein and when stored in a specialized STEP as a STEN it is stored as Dark Matter. The calculated force of it is in Einstein's gravitational equation. The cosmological constant is equivalent to an energy density in a vacuum. By equating this density to the density of the zero-point energy left in a volume after you remove all its particles by an actual experiment, you obtain a number that is 123 orders of magnitude (10^{123}) lower than what would be calculated by the gravitational equation.

Such a high value by calculation would result in a universe that would rapidly inflate, thus galaxies, would in-turn, have no time to form. This is the problem with the cosmological constant. I hypothesize that this discrepancy can be corrected if one only calculates the degrees of freedom of the surface of the vacuum STEPs instead of all STEPs in the OU.

The release of dark energy from dark matter in the STEN increases the number of STEPs that expand the OU. Currently there are ~ 10^{183} STEPs which will increase over time to ~10^{184} and eventually to 10^{500} causing the universe to convert to the universal library that will locate in a singularity or two for Souls to access.

The Kiss

The Zero Point Energy as discussed can be explained as a kiss between two people.

To blow a kiss, one would simply suck in and create a vacuum in the mouth followed by the release of the vacuum and an exhale of your breath.

The vacuum you created was actually the force of dark energy acting against the inner walls of your lips.

Now if two people engage in a mouth to mouth kiss and exchange vacuum leverage then a zero-point dark energy has been established between the two people but is not felt anywhere else.

Notes

Physicist Michio Kaku and others ask in a video (How the universe Works);

What was there before the big bang?

What banged?

Where did it bang?

 When did it bang?

Why did it bang?

Who banged it'?

"I want to know God's thoughts; the rest are details."

 My answers are as follows;

I.   What was before the Big Bang? ----------------------------infinite singularities

II. What banged? --two singularities

III. Where did it bang?   --------------------------------in the center of our universe

IV. When did it bang? ------------------------------------ the first Planck tic of time

V. Why?   -- ----------------------------so Souls could have a place to make choices

VI. Who? -------------------God, the creator of the Grand Original Design (G.O.D.)

Notes:

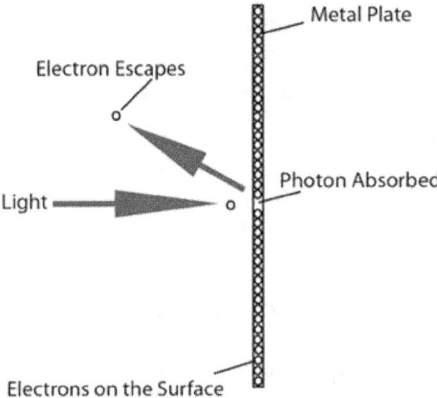

Noble Prize concepts that Einstein got and should have gotten nominated for.

1.   Special Relativity

2.   General Relativity

3.   Browning motion

4.   Quantum theory

5.   Cosmological constant & Zero Point Energy (ZPE)

6.   EPR Paradox's incomplete quantum theory

7.   Bose-Einstein condensate (BEC)

8. $E = MC^2$

9.   Proposed a Cosmic Deity

10.   The Nobel Prize in Physics 1921 was awarded to Albert Einstein "for his services to Theoretical Physics, and especially for his discovery of the law of the photoelectric effect."

Noble Prize concepts that are theorized here.

1.   Duality of the wave/particle concept mechanism.

2.   Physical properties of the M theory's 11 dimensions plus defined shapes for alternating Calabi-Yau geometry.

3.   What Dark Energy (De) is. What Dark Matter (Dm) is, and the Correlation of Dm and DE by De= DmC^2.

4.   Neutron/Proton theory for the proliferation of matter.

5.   Spooky action at a distance by synchronized STEPs.

6.   Physical properties of the spin (twist) of matter particles and wormholes of space-time.

7.   Hypothesis of Gravitational waves and graviton operation in the onion skin region of a STEP.

8.   The Physical properties of the cosmological constant and is an answer to the flatness problem.

9.   Eternal force splits off forces to manifest EM, strong, weak and gravitational forces plus the Higgs field. The Holy Grail Equation of Physics, *[Ef=DmΛ_a]* $T,$ Θ incorporates all mathematical equations know.

10.   The Grand Original Design, G.O.D. as a complete theory of everything, TOE.

Copyright of SXS glossary by Michael Boyle

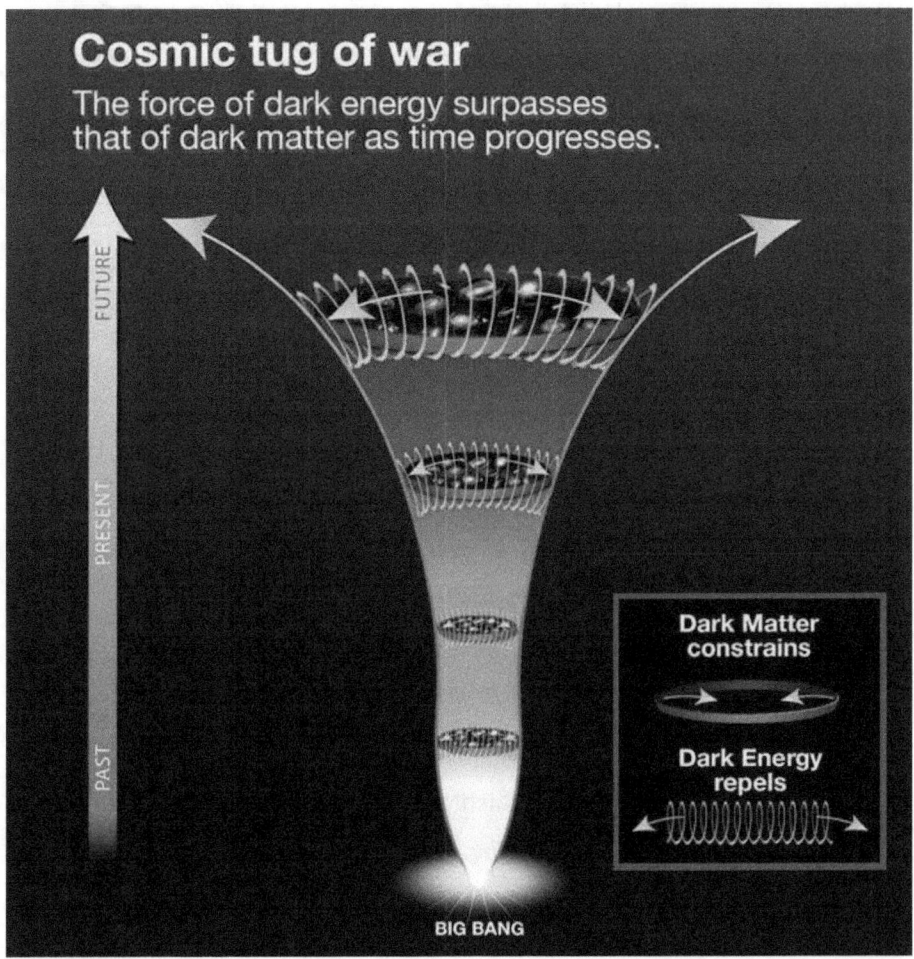

A NASA Image of the Dark Matter and Dark Energy Relationship. The G.O.D. theory of everything (TOE) explains the mechanism of how the two are similarly related to Einstein's relationship of $E = mC^2$ for normal mass and energy. Here I postulate that $De = DmC^2$.

Information about this glossary:

The text on this site was written by Michael Boyle, a member of the SXS group, currently at Cornell.

This site was made possible by the generous support of Michael Scott.

The SXS project is supported by the Sherman Fairchild Foundation, the NSF, and NASA. **http://www.caltech.edu/...http://www. cornell.edu/**

The use of this glossary does not endorse in any way the ideas, conclusions or anything that the SXS group stands for

A

Acceleration

Any change in the speed with which an object moves, or the direction in which it moves.

Active Galactic Nucleus

The central portion of a galaxy which gives off unusually large amounts of energy. These are thought to be powered by super massive Black Holes.

Amplitude

The height of the peak of a wave, measured relative to its center. Equivalently, the depth of the trough of a wave.

Antiparticle

Essentially, the "opposite" of a particle. Every type of matter has a corresponding antiparticle, with the same mass but opposite charge, for example. Other numbers describing the particle will be reversed for the antiparticle.

Atomic Nucleus See Nucleus below.

B

Big Bang

An astrophysical theory of the beginning of the Universe. It suggests that the Universe began in a very tiny region of space, and exploded outward. Astrophysicists believe that this occurred roughly 14 billion years ago. Other astrophysical theories for the beginning of the Universe—like the Braneworld theory—exist, though none is as thoroughly studied and supported by the data as the Big Bang model. Scientists have no idea what came before the Big Bang.

Big Crunch

Essentially the opposite of the Big Bang, the Big Crunch is one possible fate of the Universe. If the matter and energy of the

Universe are not moving outward quickly enough, gravity could pull the Universe in on itself, collapsing it in a final Big Crunch. It is not yet known whether this will happen to our Universe.

Black Hole

A region of space-time where gravity is so intense that nothing—even light—can ever escape. Objects may fall in to the Black Hole, but once they pass the Event Horizon, they can never escape again. Most Black Holes believed to exist are thought to be formed in the collapse of very large stars, or the collision of stars or other Black Holes.

Boson

A type of particle with "integral angular momentum"—a spin of 0, 1, 2, etc. Spin refers to an intrinsic quality of all particles. Examples of Bosons are photons (which are the particles which give us light) and graviton (which give us gravity). The other type of particle is the fermion.

Boundary division between two regions. Physicists frequently analyze only a part of a larger system, so that they do not need to keep track of everything. This usually simplifies the analysis, but requires an understanding of what happens at the boundary. Interesting computer simulations usually require a boundary. Boundary Conditions

The state of a physical system at a boundary. Interesting computer simulations usually require boundary conditions.

Brane

Objects which arise in string theory. They can have any number of dimensions, and are usually imagined as existing in a space with more dimensions than the brane itself has.

Braneworld

A four-dimensional surface—a "brane"—in a space-time with more than four dimensions.

C

Causality

The relationship between cause and effect. Typically, we assume that a given event is a result of events that came before it in time.

Compact Binary

A specific type of Binary system in which both members are compact (meaning they are White Dwarfs, Neutron Stars, or Black Holes) and have roughly equal mass.

Cosmic String

A long, heavy object from Quantum Field Theory or String Theory, which is very thin. They may have been created in the early life of the Universe, and would now stretch across the entire Universe. No cosmic string has ever been observed. They may or may not exist.

Cosmological Constant

A mathematical device used by Einstein to keep the Universe from falling in on itself. He later called this his "greatest blunder", because it kept him from predicting the expansion of the Universe. Astrophysicists now believe there may be a use for the cosmological constant.

D

Dark Matter

A type of matter which is found near other matter. It cannot be observed directly (it is "dark"), but can be noticed as a result of the pull of gravity from the dark matter. Astrophysicists believe that more than half of the Universe could be in the form of dark matter.

Differential Geometry

The branch of geometry which deals with curved surfaces and spaces. Calculus is used to analyze the shape of these surfaces and spaces. Differential geometry is a key tool used in the study of General Relativity.

E

Einstein's Equations

A set of "tensor" equations Einstein devised to describe how mass warps space-time. The set of equations may be written as $G = 8\pi T$, where both G and T each represent a set of ten quantities. The G quantities represent the warping of space-time, while the T quantities—the "Stress-Energy tensor"—represent the mass.

Electromagnetic Wave

An electric and magnetic disturbance that travels through space like a wave. What we experience as light is an electromagnetic wave. Electromagnetic waves therefore travel at the Speed of Light. Other types of electromagnetic wave range from Radio Waves and Microwaves, through to X-Rays and Gamma Rays.

Electron

A tiny particle usually found swirling around an atomic nucleus, the electron carries the standard unit of negative charge, which balances the positive charges in a nucleus. The interaction between electrons and nuclei is responsible for chemistry. Electrons can become detached from the nucleus, when given enough energy, and become free. Electrons are members of the particle class called fermions, and are roughly 2,000 times lighter than neutrons and protons.

Elementary Particle

A particle which is not made up of smaller particles. The neutron and proton are both made up of smaller particles—called quarks. On the other hand, quarks are not believed to be made of anything smaller. Like quarks, electrons and photons are also believed to be elementary particles.

EMRI

See Extreme Mass-Ratio Inspiral below

Epicycle

A secondary circle centered on another, usually larger, circle. Its center moves along the edge of the main circle.

Event Horizon

A surface—like the one surrounding a Black Hole—enclosing a region of space from which nothing (even light) can ever escape.

Exclusion Principle

A rule in Physics which says that no two identical particles can be in the same state (position, for instance) at the same time. This principle only applies to fermions, not to bosons. It is usually referred to as the "Pauli Exclusion Principle", after its inventor Wolfgang Pauli.

Extreme Mass-Ratio Inspiral

A particular type of binary in which there is a very large difference in the masses of the two objects. Generally, this will involve a super- massive Black Hole with a mass millions of times that of our Sun, and a Neutron Star or Black Hole with a mass roughly the same as our Sun.

F

Failure of Simultaneity

Simultaneous events are events in different places which happen at the same time. It turns out that this concept depends on how quickly one is moving. That is, if two observers are moving relative to each other, they will not be able to agree on the simultaneity of events.

This is the Failure of Simultaneity.

Fermion

A type of particle with "odd half-integral angular momentum"—a spin of 1/2, 3/2, etc. Spin refers to an intrinsic quality of all particles.

Examples of fermions are electrons, neutrons, and protons. The other type of particle is the boson.

First Law of Motion

The first of Newton's Laws of Motion, which says that moving objects move in a straight line. Specifically, the Law says, "An object at rest tends to stay at rest and an object in motion tends to stay in motion with the same speed and in the same direction unless acted upon by an unbalanced force."

Fission

See "Nuclear Fission".

Flatness Problem

The unexpected result that the Universe is not expanding so slowly that it will clearly collapse back on itself in a Big Crunch, nor expanding so quickly that it will clearly keep expanding forever. Instead, measurements show that the Universe is treading a fine line between the two—the Universe is referred to as "flat". Astrophysicists would not expect this to be the case unless there were some cause for the Universe to tread such a fine line. Finding this cause is the "flatness problem".

Frequency

The number of occurrences of something in a given period of time. For a wave, this is might be the number of times the wave peaks in one second.

Fusion

See "Nuclear Fusion"

G

General Theory of Relativity

Einstein's version of the laws of physics, when there is gravity. Building on the Special Theory of Relativity, this theory generalizes Einstein's work so that the laws of physics must be the same for all observers, even in gravity. Einstein showed that gravity is best understood as a warping of the geometry of space-time, rather than as a pulling of objects on each other. The crucial idea is that object move along geodesics—which are determined by the warping of space-time—while space-time is warped by massive objects according to the formula $G = 8 \pi T$.

Geodesic

Essentially the "straightest path" is a curved space or curved space-time. This is the path followed by an object with no forces acting on it. In the curved space-time of General Relativity, these paths may seem to be much curved—even appearing as circles or ellipses, for example. A geodesic is easily understood by looking at a very small region around the object. Even in highly curved space-time, a small enough region will seem flat, so there is a natural idea of a "straight path". By following short segments, the whole geodesic is built up into one long path.

Gravitational Wave

A gravitational disturbance that travels through space like a wave. This type of wave is analogous to an Electromagnetic Wave.

Gravitational waves are given off by most movements of anything with mass. Usually, however, they are quite difficult to detect.

Physicists are currently working hard to directly detect gravitational waves. Experiments like LIGO and LISA are designed for this purpose.

H

Horizon Problem

A problem with the simplistic Big Bang theory having to do with the smoothness of the Universe. The Early Universe should have been very random in terms of the temperature and density of different parts of space. This randomness should not have had time to distribute itself more evenly. Yet, this is what is observed in the Universe. The theory of Inflation solves this problem.

Inflation

A brief period shortly after the Big Bang during which the Universe expanded very rapidly. The theory of Inflation is necessary to make the theory of the Big Bang agree with astronomical observations.

Initial Data

Physicists frequently analyze a physical situation starting at some moment in time, and ignore what happened before that moment in time. To do this, however, they need to understand what was happening at that initial moment. That is, they need Initial Data. This is very similar to using Boundaries, which requires having Boundary Conditions.

Inspiral

The gradually-shrinking orbit of a binary system. As the pair of stars in the binary orbit each other, they give off energy in the form of gravitational waves. This lost energy draws them closer in their orbit—eventually resulting in a Merger.

Interference

A phenomenon which can occur any time there is any type of wave, which amounts to two waves canceling each other out. If two different waves meet at the same place, and one would hit its peak while the other would hit its trough, the waves will cancel, and there will be no disturbance. Alternatively, if both waves would hit their peaks at the same time, the waves will boost each other, so that there is a greater disturbance.

Interferometer

A scientific device which makes use of the Interference of waves—typically, light waves. This type of device can measure changes in length with extraordinary precision, and forms the basis of modern gravitational wave detectors.

L

Light Year

The distance traveled by light in one year. This is roughly 10^{13} kilometers, or 6×10^{12} miles.

M

Merger

The portion of the Inspiral of a binary system in which the individual objects are highly distorted, and their orbit is changing rapidly.

This portion is not well-understood, and must be simulated using Numerical Relativity.

Metric

A set of numbers which encodes information about the geometry of space or space-time. Together with an understanding of how that information is encoded (using Coordinates) everything about the geometry can be understood.

N

Neutrino

A type of particle which has no charge and an extremely small mass. It is a Fermion, and is extremely difficult to stop or to detect.

Nonetheless, they are produced in large numbers. The Sun, for example, sends 30 million neutrinos through every square inch of the Earth every single second. They are so hard to stop, however, that if a neutrino were sent through a solid light year of lead, it would still have a 50:50 chance of flying right through without stopping.

Neutron

One of the particles in an atomic nucleus. These particles have no electric charge, but they hold together the protons (positive particles in a nucleus), and account for roughly half of the particles in the nucleus. Neutrons are fermions, and are believed to form the majority of the matter in a neutron star.

Neutron Star

A type of star which is very old, having cooled off and stopped nuclear fusion reactions. When gravity pulls the star down on itself, the electrons and protons are squeezed together, leaving just neutrons. The star is then supported against gravity by "neutron degeneracy pressure" (no two neutrons can be in the same place at the same time). These are produced when a star is too heavy to be a white dwarf, but not heavy enough to turn into a Black Hole.

Newton's First Law of Motion See "First Law of Motion".

Nuclear Fission

A physical process which takes a nucleus of a heavy element (like Uranium or Plutonium, for example) and breaks them down into two or more smaller nuclei. This process releases large amounts of energy. It is sometimes called "Splitting the Atom". This is used in many modern power plants to generate electricity by heating water with the energy released. The two or more smaller nuclei which are produced are frequently toxic, and nearly always radioactive, which makes this method of producing electricity controversial. Fission is also the method used in simple nuclear weapons.

Nuclear Fusion

A physical process which takes light elements and combines them into a heavier element. An example is the fusion of two Hydrogen atoms to form a Helium atom. This is the process which gives stars (including our Sun) heat and energy so that they shine. Though fusion is not yet used to produce electricity, scientists are working on this method, which will provide a nearly-inexhaustible source of cheap electrical energy, with little of the pollution or danger of nuclear fission. Fusion is also the force behind some very powerful nuclear weapons—the "H-Bomb". Understand things like the merger of two Black Holes.

Nucleus

The central part of an atom, which contains Neutrons and Protons. Electrons are usually found around the Nucleus. Strictly speaking, this is the only part of an atom involved in Nuclear Reactions (Fission or Fusion).

Numerical Relativity

The branch of Relativity research which deals with simulating the development of Space-time, using computers. This is believed to be the only possible way to understand things like the merger of two Black Holes.

O

Observer

A person or piece of equipment that measures something in physics. Frequently, we speak of an observer measuring time or a distance in a particular place.

P

Parsec

A measure of distance which is roughly 3.26 Light Years. Pauli Exclusion Principle

See Exclusion Principle, above.

Period

The length of time between two events. For a wave, this is usually the length of time it takes two successive peaks to pass a given point. This number is simply 1 divided by the Frequency of the wave.

Phase

1. For a wave, the position of any particular feature of the wave.

2. For matter, a distinct form of a substance, such as solid, liquid, or vapor.

Phase Transition

A change of the state of matter from one phase to another, such as the transition from liquid water to solid ice.

Photon

An Elementary Particle which carries the energy of light. The photon is a Boson, and has no mass. It always moves at the Speed of Light

Pi, π

(Pronounced as "pie".) An important number in geometry. This is defined to be the ratio of the circumference of a circle to the diameter of that circle, in flat space. π is an irrational number, which means that its exact value cannot be written down, though it can be calculated as precisely as necessary. Its value is approximately 3.14159265358979323...*In my concept of the OU, pie would end at the 10^{500th} didget after the decimal.*

Powers of 10

In order to write very large, or very small numbers efficiently, scientists use a system of writing in powers of 10. For example, the number 108 represents 10 multiplied by itself 8 times, which is a 1 with 8 zeroes after it: 100,000,000. Thus, rather than writing

300,000,000, a scientist will write 3 × 108. Similarly, a number like 10-21 represents 0.1 multiplied by itself 21 times, which is a 1 with 21 zeroes in front of it (including the one before the decimal point): 0.000000000000000000001.

Proton

One of the particles in an atomic nucleus. These are electrically positive particles which attract electrons to the atom. Protons are fermions, and are very similar to neutrons, except that they have electric charge, and a slightly higher mass.

Pulsar

A neutron star with a very high rate of spin, and very intense magnetic fields. The pulsar gives off beams of radiation along its magnetic poles. If these poles are not aligned with the spin poles, the beam will sweep around like the beam of a lighthouse.

Q

Quantum Field Theory in Curved Space-time

A theory which attempts to incorporate ideas from Quantum Mechanics and General Relativity. This theory is needed when the quantum behavior of objects is important, and there is extreme curvature of space-time.

Quantum Mechanics

A modern physical theory which is vital to describing extremely small objects, like electrons around an atom. Though this theory also applies to larger objects, its effects become very similar to those of Newton's theory—which is typically much easier to use and understand. One of the most important ideas in Quantum Mechanics is the Uncertainty Principle.

Quark

An Elementary Particle which makes up the Neutron and the Proton, as well as many more exotic particles.

These particles are Fermions, and have charges of either 2/3 as much as an electron's charge, or -1/3. They come in "flavors" of up, down, strange, charmed, bottom, and top, as well is in "colors" of red green and blue. The "flavor" and "color" are just fanciful names given to describe intrinsic properties of these particles—similar to charge, mass, or spin.

Quasar

A very dense and very bright object seen in the distant Universe. Quasars are believed to be powered by Black Holes.

R

Randall-Sunder Model

A theory of the Universe which postulates that we live in a five-dimensional space-time, though we are confined to a four- dimensional Slice of it. This model is still being explored as an accurate description of the Universe in which we live.

Rundown

The portion of an Inspiral, following the Merger, when the two objects have combined into one. During this brief period, the combined object will settle down by giving off gravitational waves.

S

Singularity

A point at which the curvature of space-time becomes infinite. Singularities can form—for example—when too much matter is squeezed into a region which is too small. Singularities are found at the center of a Black Hole, at the beginning of the Universe in the Big Bang, and at the end of the Universe (if it ever comes) in the Big Crunch. Many scientists believe— though there is no solid evidence—that all singularities in the present Universe lie hidden behind Event Horizons. This is the "Cosmic Censorship Conjecture"

Singularity Theorem

A theorem (a mathematically proven fact) which shows that a singularity must exist under certain circumstances. For example, physicists have shown that the Universe must have begun with a singularity.

Space-Time

A concept in physics which merges our usual notion of space with our usual notion of time. Just as space as we know it has three dimensions, space-time has four dimensions—with time as the fourth dimension.

Special Theory of Relativity

Einstein's version of the laws of physics, when there is no gravity. The two fundamental concepts in the foundation of this theory are equality of observers, and the constancy of the speed of light. The first of these means that the laws of physics must be the same, no matter how quickly an observer is moving. The second means that everyone measures the exact same speed of light. This theory is useful whenever the effects of gravity can be ignored, but objects are moving at nearly the speed of light. It has been successfully tested many times in particle accelerators, and orbiting spacecraft. For objects moving much more slowly than light, Special Relativity becomes very nearly the same as Newton's theory, which is much easier to use.

Speed

Speed of Light

A constant of Nature. This speed is precisely 299,792,458 meters per second, or roughly 670,616,629 miles per hour. One of the most unusual discoveries of science has been the fact that all Observers measure light as moving at exactly this speed, even if those observers are moving relative to each other. This fact is one of the basic ingredients in Einstein's Special Theory of Relativity.

Spin

An intrinsic property of particles. (That is, a property which does not change. Mass and electric charge are examples of intrinsic properties.) Spin is related to the usual notion of spin, though it is a little unusual. Spin comes in units of 1/2, so that a particle may have a spin of 0, 1/2, 1, 3/2, and so on. A particle's spin determines whether it is a Fermion or a Boson.

Standing wave

Standing waves occur in confined spaces – in a microwave oven, on a violin string. They do not transport energy from one place to another.

Pertaining to a process involving a randomly-determined sequence of observations.

String

The fundamental object in String Theory, which replaces the notion of a particle in standard Quantum Mechanics. Rather than being a simple point-like object, fundamental particles become tiny strings or loops. The vibrations of these strings result in various properties like spin.

String Theory

A theory of physics taking the String as its fundamental object. This theory attempts to solve problems in standard Quantum Mechanics and Quantum Field Theory. It actually predicts the existence of gravity.

Superseding

Another name for a "Super-symmetric String", which is usually referred to simply as a String.

Supernova

Violently exploding stars which shine very brightly for days or weeks. They occur when the fuel for nuclear reactions is used up, and a star cools. Gravity pulls all the matter down toward the star's center. If this happens quickly, nuclear reactions may suddenly begin again, detonating the star in a nuclear explosion.

T

Traveling wave

Traveling waves are observed when a wave is not confined to a given space along the medium. The most commonly observed traveling wave is an ocean wave.

U

Uncertainty Principle

The principle of Quantum Mechanics—as well as Quantum Field Theory and String Theory—which says that an observer can never know both the position and velocity of a particle with perfect precision. Specifically, the more certain an observer is of the position, the less certain that observer must be of the velocity, and vice-versa.

V

Vacuum Energy

The energy that is present even in otherwise empty space. This energy has been measured to exist (in the "Casimir Effect").

Whereas matter causes the expansion of the Universe to slow down, vacuum energy actually causes the expansion to speed up.

Virtual Particle

A particle which cannot be directly detected, but is assumed to exist due to its indirect effects on real particles. Virtual particles come in pairs from the vacuum of space.

W

Wavelength

The distance between two neighboring peaks or troughs of a wave.

White Dwarf

A type of star which is very old, having cooled off and stopped nuclear fusion reactions. A white dwarf is supported by "electron degeneracy pressure" (no two electrons can be in the same place at the same time). These are produced when a star is not heavy enough to turn into a Neutron Star or a Black Hole.

Worm Hole

A tube in space-time which connects two widely separated places in the Universe. Wormholes could provide the possibility of time   travel, though they would probably be very unstable, and only exist for a short period of time.

X

X-Ray

A type of light—or electromagnetic wave—which is invisible to the naked eye. X-Rays are much more energetic than the light we see. They can penetrate skin very easily, for example. In the doctor's or dentist's office, X-Rays are detected on a photographic plate, allowing us to see inside the body. Like all other forms of electromagnetic radiation, X-Rays travel at roughly 300,000,000 meters per second (186,000 miles per second).

Y Z

Postface

My preface for this book was to show how a dyslexic condition can relate to creative thinking.

Also, I wanted to challenge learned scholars to prove or disprove any or all of my thought experiments.

I in no way think that I could tie Einstein's shoe laces. My thoughts are based on his view of quantum theory etc.

Everything that I have postulated in this book may not bear fruit, but just maybe the underlying concepts will.

A paradigm shift away from atheistic theoretical physics concepts to a more deity-agnostic or spiritual approach is needed.

The only way a synchronized STEP arrangement of the universe is possible, is if our Creator designed it, since a statistical random approach would not be possible. Each synchronized movement of any of the 10^{183} STEPs in our observable universe is not random. It is by choice!

 The cosmological constant said to be the "worst prediction in physics", when calculated by quantum volume is off by 10^{120} orders of magnitude. However, when quantum theory is entangled with $\sim 10^{120}$ STEP surface areas of the OU, it is in better agreement with density calculations "here-to-for."

This is an example of a more complete theory, as suggested by Einstein in the EPR publications.

My "Holy Grail Equation", $\boldsymbol{Ef = Dm\Lambda_a}$ is the under lying mathematics of the theory of everything (TOE) denoted as the "Grand Original Design".

Great minds like Howard Georgi, Leonard Susskind, Anthony Lisi Garrett, Lee Smolin, Stephen Hawking, Lisa Randall, Edward Witten, Anton Zelinger

Michio Kaku, etc. are excellent candidates to explore my STEP approach to the Grand Original Design, G.O.D.

Some of these scientist are atheists, some are not, but if we all keep an open mind, much will be accomplished.

Oprah Winfrey and her spiritual peers, such as Gary Zukav, Deepak Chopra and Paulo Coelho etc., are excellent candidates to critically review spiritual concepts developed from my Grand Original Design (G.O.D.) Theory of Everything (T.O.E.).

The phrase "When you want something, the whole world will help you" written by Paulo Coelho in his book, "The Alchemist" is supported by STEP synchronization. It might just turn out that the synchronization of STEPs is the mechanism of how the whole world can help you.

The Observable Universe (OU) is to Eternity as Finite is to Infinity, Finite is to Atheist as Spiritualism is to Eternity.

The OU is Finite and Eternity is Infinite. The transition from live to dead is the transformation from a finite life back to our infinite eternal being. Atheists do not believe this transformation and hold their beliefs in the finite. The finite of infinity is a wonderfully place, but it is not the end.

The universe was created so Souls could make choices.

Currently Islamic terrorists' religious wars and racial prejudice are very poor choices, while the pursuit for liberty, justice and knowledge are good noble choices.

The subconscious of the OU is the synchronizations of the STEPs and the consciousness of the OU is Mother Nature. The subconscious of a human is its soul and consciousness is making choices.

Vector- Determinacy. (NEW): the forward momentum of matter through space-time (with non-local Time-Channels) to create an effect that is produced by the cause of its purpose.

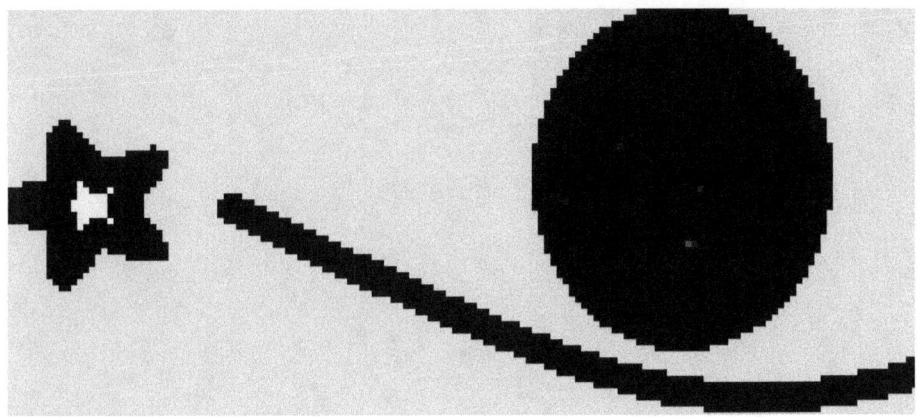

The eclipse of the sun by the moon to show that Einstein's bending of light was a gigantic revelation in the early part of the 20th century. Einstein described it as matter bending the fabric of space-time. This is the way Einstein explained how gravity works. He was right in all aspects of his conclusions.

I only add to his synopsis that this warping of space-time at the Planck's level is the passage through it by elementary particles. This passage is by tesseract rotation and time channel superluminal(modified) travel by a traveling wave with standing wave disks. The travel is by **Vector-Determinacy**. The purpose of the photon(s) from the star is to shine as a vector (momentum x velocity) in a determined direction. The cause of the shining is due to the escaping of the photon(s) from the surface of the star that travel directionally. The effect of the cause is the way in which it travels and how it is received. Like the double slit experiment; When obstacles are in the way of direct passage, elementary particles will find a way **(Vector- Determinacy)** to arrive to a location that is within its standing wave disk.

It's not the STEPs being warped, it's the vector-determinacy that is causing the warp. Warping now can mean that the space-time fabric is merely the tesseract rotating of the individual STEPs, coupled with time channel superluminal(limited) involvement.

Note; also the picture above shows the warping to be increments of cube arrangement (not to scale) representing STEPs that elementary particles pass through.

www.ingramcontent.com/pod-product-compliance
Lightning Source LLC
Chambersburg PA
CBHW070852180526
45168CB00005B/1788